A Programmed Approach To
The Circulatory System

by

George I. Sackheim

ISBN 978-0-87563-014-4

Published by

Stipes Publishing L.L.C.
204 W. University Ave.
Champaign, Illinois 61820

TABLE OF CONTENTS

DIRECTIONS

Place a blank sheet of paper over the following page and move it downward to expose the first statement or question as indicated by the number 1 in the upper left-hand corner of the frame. If the first frame contains a statement, read it carefully and proceed to the next frame. If the frame contains a question, answer that question and then move the blank paper down to expose the correct answer. If your answer was correct, proceed to the next frame, uncovering only the question. Answer it and then compare the answer with your own. If you did not answer the question correctly, reread the question and try it again. When you come to the bottom of a page, transfer the blank paper to the following page and continue as before.

I. THE HEART

1. The purpose of this part of the program is to trace the path of the blood from the time it enters the heart to the time it returns to the heart to again pass through the same cycle.

---

2. Blood entering the heart flows into the right atrium.

---

3. When blood flows into the heart, it first enters the right ______________.

atrium

---

4. The _________ atrium is the part of the heart which first receives blood.

right

---

5. Blood entering the heart flows into the ______________________.

right atrium

---

6. When blood enters the heart, it enters the right atrium as indicated in the following diagram:

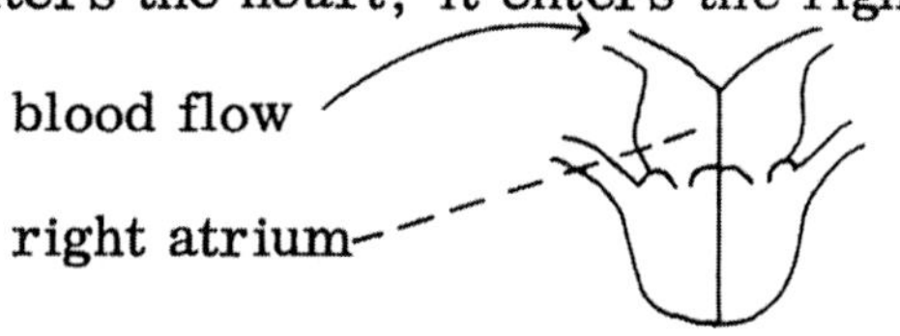

---

7. From the right atrium the blood passes through a valve known as the tricuspid valve.

---

8. From the right atrium the blood flows through the tri________ valve.

cuspid

---

9. Blood entering the heart flows into the ______________________.

right atrium

---

10. The valve by which blood leaves the right atrium is known as the _________________ valve.

tricuspid

---

11. The blood flows out of the right atrium through the _______________.

tricuspid valve

---

12. In the following diagram, label the right atrium.

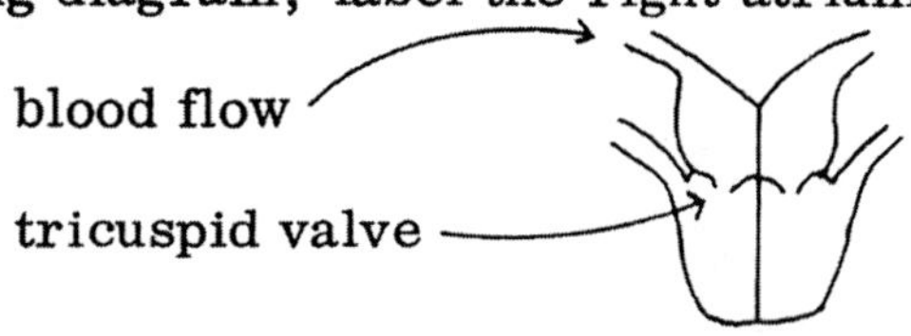

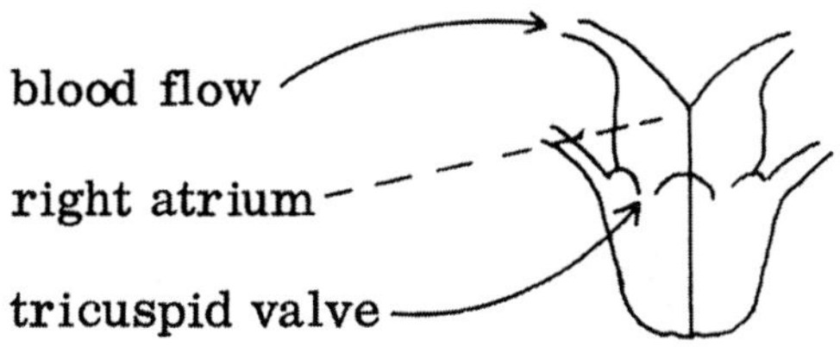

1401. From the femoral vein the blood flows into the ________________ vein

external iliac

1402. and then into the ____________________ vein.

common iliac

1403. From the common iliac vein the blood flows back to the heart through the ______________.

inferior vena cava

1404. **Trace the flow of the blood from the heart to the right leg and return to the heart.**

Heart to ascending aorta to arch of aorta to thoracic aorta to abdominal aorta to right common iliac artery to external iliac artery to femoral artery to popliteal artery to anterior and posterior tibial arteries. From anterior tibial artery to dorsalis pedis artery to plantar arch of foot. From posterior tibial artery to peroneal artery and also to medial and lateral plantar arteries to plantar arch of foot. From plantar arch of foot to arterioles, capillaries and venules to dorsal venous arch and plantar venous arch. From dorsal venous arch superficially into small and great saphenous veins. From small saphenous vein into popliteal vein to femoral vein and from great saphenous vein to femoral vein. From dorsal venous arch to deep veins through dorsalis pedis vein to anterior tibial vein to popliteal vein to femoral vein. From plantar venous arch to medial and lateral plantar veins to posterior tibial vein to popliteal vein to femoral vein. From femoral vein to external iliac vein to common iliac vein to inferior vena cava to heart.

13. The blood flows through the tricuspid valve into the right ventricle.

---

14. After the blood passes through the tricuspid valve, it enters the ______________ ventricle.

right

---

15. The right ______________ receives the blood passing through the tricuspid valve.

ventricle

---

16. Blood enters the heart at the ________________________.

right atrium

---

17. From the right atrium the blood passes through the ____________ valve,

tricuspid

---

18. and enters the ___________________.

right ventricle

---

19. Label the indicated parts:

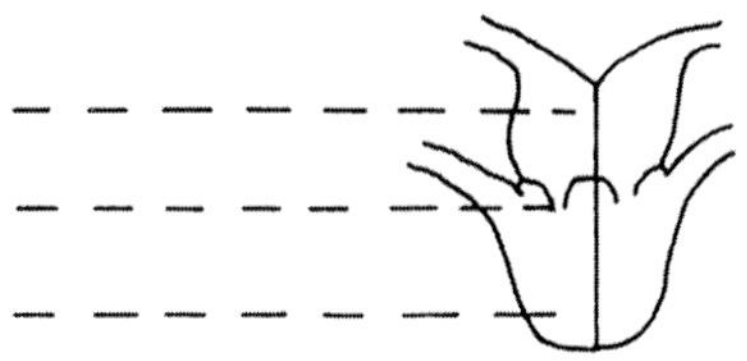

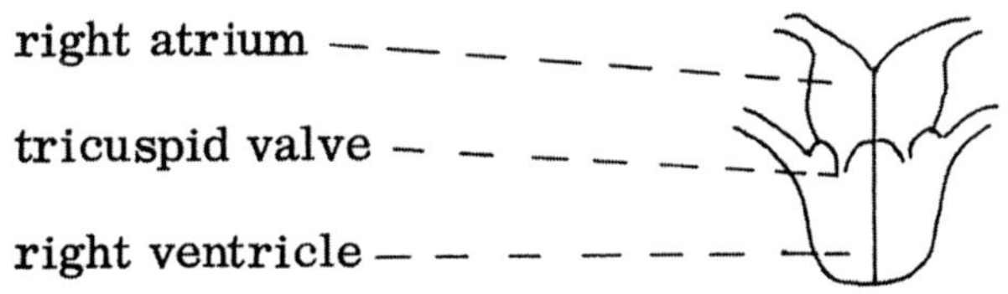

---

20. From the right ventricle the blood passes through the semilunar valves, (semi means half and lunar refers to the moon, so they are half-moon shaped valves).

---

21. The valves by which the blood leaves the right ventricle are called the semi___________ valves.

lunar

---

22. The _________ lunar valves are the valves by which the blood flows out of the right ventricle.

semi

---

23. From the right ventricle the blood flows out through the ____________________ valves.

semilunar

1386. and then into the ________________ arteries

---

medial and lateral plantar

1387. which join to form the ________________ of the foot.

---

plantar arch

1388. From the plantar arch of the foot the blood flows through arterioles, capillaries, venules and then returns through the ____________ arch and the ____________ arch.

---

dorsal venous, plantar venous

1389. From the dorsal venous arch the blood flows superficially into the ____________ veins.

---

small and great saphenous

1390. The small saphenous vein empties into the ________________ vein

---

popliteal

1391. which in turn empties into the ____________ vein.

---

femoral

1392. The great saphenous vein empties into the ____________ vein.

---

femoral

1393. From the dorsal venous arch the blood flows to the deep veins through the ____________ vein

---

dorsalis pedis

1394. which extends into the ________________ vein.

---

anterior tibial

1395. The anterior tibial vein empties into the ________________ vein

---

popliteal

1396. which then leads to the ______________ vein.

---

femoral

1397. From the plantar venous arch the blood flows into the ______________ veins

---

medial and lateral plantar

1398. which carry the blood to the ______________ vein.

---

posterior tibial

1399. From the posterior tibial vein the blood flows into the ______________ vein

---

popliteal

1400. which empties into the ______________ vein.

---

femoral

24. Label the following diagram:

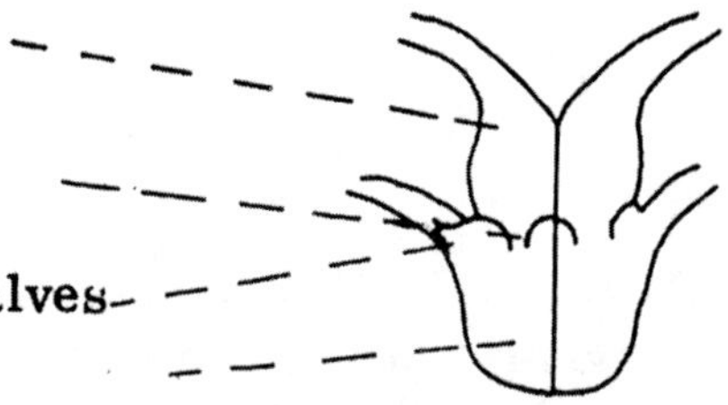

semilunar valves

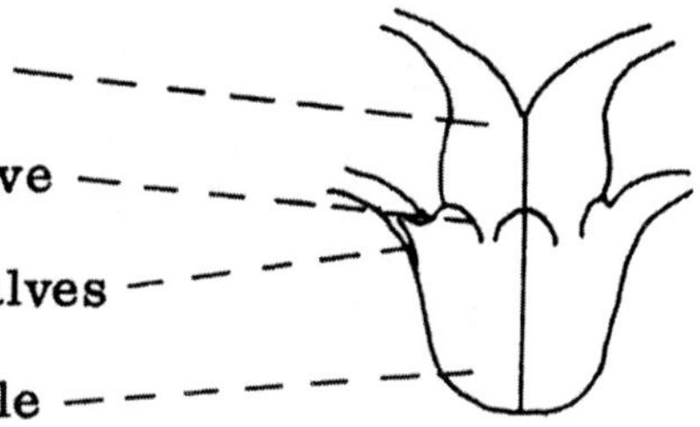

25. Reviewing, blood enters the heart at the ________________,

right atrium

26. then flows through the ______________ valve

tricuspid

27. and enters the __________________.

right ventricle

28. From the right ventricle the blood flows into the __________________ valves.

semilunar

29. After passing through the semilunar valves, the blood enters the pulmonary artery.

30. From the semilunar valves the blood enters the ______________ artery.

pulmonary

31. The pulmonary _____________ receives the blood after it passes through the semilunar valves.

artery

32. The blood flows from the right ventricle through the semilunar valves to the ____________________________.

pulmonary artery

1371. From the external iliac vein the blood flows into the ________________ vein.

common iliac

1372. The common iliac veins (right and left) empty into the inferior vena cava which then carries the blood back to the heart.

1373. From the common iliac veins the blood flows into the ______________ and then back to the heart.

inferior vena cava

1374. Time for a complete review of the circulation to the lower extremities.
The blood flows from the heart into the ______________ aorta

ascending

1375. into the ____________

arch of the aorta

1376. into the ________________ aorta

thoracic

1377. and then into the ______________ aorta.

abdominal

1378. At the terminal end of the abdominal aorta the blood flows into the (right and left) ______________ arteries.

common iliac

1379. From the common iliac artery the blood flows into the ________________ artery

external iliac

1380. and then into the ________________ artery.

femoral

1381. From the femoral artery the blood flows into the ______________ artery

popliteal

1382. which divides into the ____________ artery and the ____________ artery.

anterior tibial, posterior tibial

1383. From the anterior tibial artery the blood flows into the ______________ artery

dorsalis pedis

1384. and then into the ______________ of the foot.

plantar arch

1385. From the posterior tibial artery the blood flows into one branch, the ____________ artery

peroneal

33. Label the indicated parts:

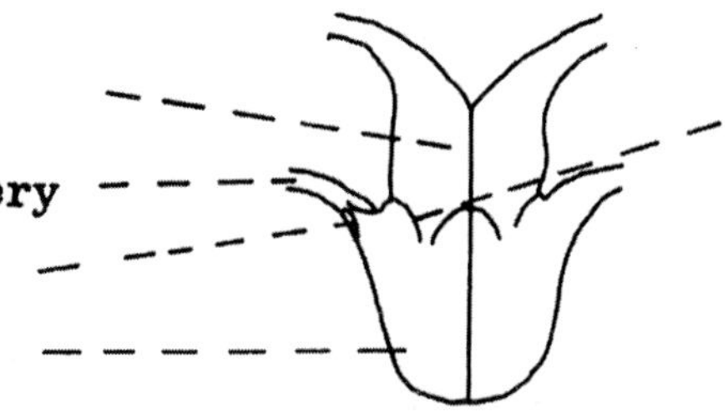

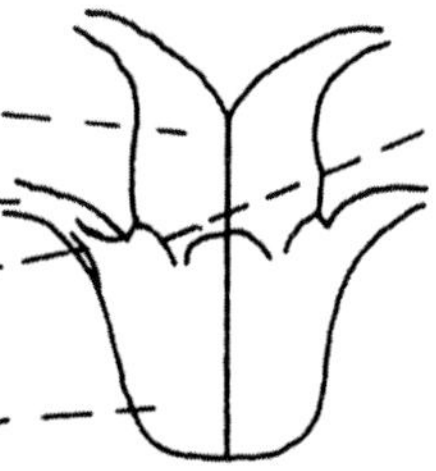

---

34. The pulmonary artery carries the blood to the lungs.

---

35. The blood travels to the __________ by means of the pulmonary artery.

lungs

---

36. The blood flows from the semilunar valves into the __________ __________ and then goes to the __________.

pulmonary artery, lungs

---

37. Reviewing again, (if you feel that you do not need a review, proceed to frame 43) blood enters the heart at the __________.

right atrium

---

38. flows through the __________

tricuspid valve

---

39. into the __________.

right ventricle

---

40. The blood then flows through the __________

semilunar valves

---

41. into the __________

pulmonary artery

---

42. which carries the blood to the __________.

lungs

1356. into the ________________ vein.

anterior tibial

1357. From the anterior tibial vein the blood flows into the ______________ vein

popliteal

1358. which empties into the ______________ vein.

femoral

1359. From the plantar venous arch the blood flows into the ________________ veins

medial and lateral plantar

1360. and then into the ____________ vein.

posterior tibial

1361. From the posterior tibial vein the blood flows into the ________________ vein

popliteal

1362. which then empties into the ______________ vein.

femoral

1363. The femoral vein empties into the <u>external iliac vein</u>.

1364. From the femoral vein the blood flows into the external ______________ vein.

iliac

1365. The ______________ iliac vein receives the blood flowing from the femoral vein.

external

1366. The blood flows from the popliteal vein into the femoral vein and then into the __________ vein.

external iliac

1367. The femoral vein empties into the external iliac vein which in turn empties into the <u>common iliac vein</u> (right common iliac vein for the right leg and left common iliac vein for the left leg).

1368. The blood flows from the external iliac vein into the _____________ iliac vein.

common

1369. The common ______________ vein receives the blood flowing from the external iliac vein.

iliac

1370. From the femoral vein the blood flows into the _____________ vein.

external iliac

43. Label the following diagram:

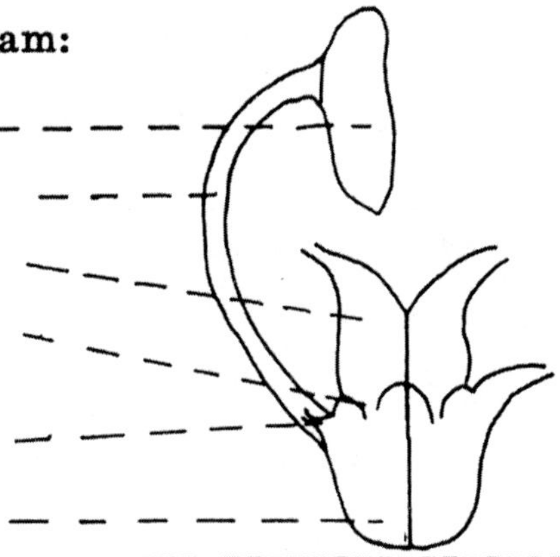

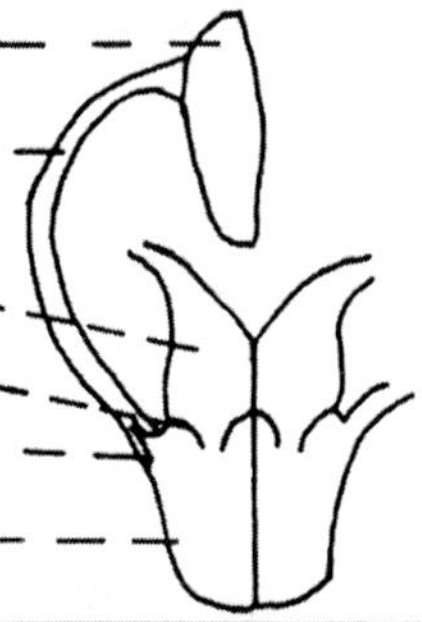

---

44. In the lungs the pulmonary artery subdivides into many tiny branches called <u>capillaries</u> (pronounced kap' il er iz)

---

45. The blood flowing through the lungs passes through many tiny ________________.

capillaries

---

46. The capillaries receive the blood from the ________________.

pulmonary artery

---

47. When the pulmonary artery subdivides, it breaks up into many ________________.

capillaries

---

48. The capillaries in the lungs are in direct contact with many tiny air sacs called alveoli (al ve' o li).

---

49. The tiny air sacs are called alv_________.

alveoli

---

50. The blood flows to the lungs, through the capillaries, around the air sacs called ______________________.

alveoli

---

51. The ______________ artery carries the blood to the lungs.

pulmonary

---

52. In the lungs the pulmonary artery subdivides into many branches called ________________.

capillaries

1342. The medial and lateral plantar veins empty into the posterior tibial vein.

1343. The posterior ______________ vein receives the blood flowing from the medial and lateral plantar veins.

tibial

1344. The blood flows from the plantar venous arch into the medial and lateral plantar veins to the ______________ vein.

posterior tibial

1345. From the posterior tibial vein the blood flows into the popliteal vein.

1346. The vein which receives the blood flowing from the posterior tibial vein is the ______________ vein.

popliteal

1347. The blood flows from the medial and lateral plantar veins into the ______________ vein

posterior tibial

1348. and then into the ______________ vein.

popliteal

1349. From the popliteal vein the blood flows, as before, into the ______________ vein.

femoral

1350. Time for another review (but if you know, then on to frame 1363 you go).
The blood returning from the foot flows from the venules into the ______________ arch and the ______________ arch.

dorsal venous, plantar venous

1351. From the dorsal venous arch the blood flows superficially into the ______________ veins.

small and great saphenous

1352. The small saphenous vein empties into the ______________ vein

popliteal

1353. which in turn empties into the ______________ vein.

femoral

1354. The great saphenous vein empties into the ______________ vein.

femoral

1355. From the dorsal venous arch the blood flows to the deep veins through the ______________ vein

dorsalis pedis

---

53. The capillaries carry the blood around the _______________.

---

alveoli

54. As the blood flows around the alveoli, it picks up oxygen and at the same time gives off carbon-dioxide.

55. As the blood flows through the lungs, it picks up _________ from the alveoli.

---

oxygen

56. As the blood flows through the lungs, it gives up ______________ to the alveoli.

---

carbon dioxide

57. The blood picks up oxygen from and gives off carbon dioxide to the _______________.

---

alveoli

58. Time for another review. (If you feel that you know the material and do not need a review, proceed to frame 68).
The blood enters the heart at the _______________,

---

right atrium

59. then flows through the ____________________

---

tricuspid valve

60. into the ____________________.

---

right ventricle

61. From the right ventricle the blood passes through the _________________

---

semilunar valves

62. into the ____________________

---

pulmonary artery

63. which carries the blood to the _______________.

---

lungs

64. In the lungs the pulmonary artery subdivides into many branches called _______________.

---

capillaries

65. These capillaries are in contact with the air sacs called _______________.

---

alveoli

66. From the alveoli the blood picks up _______________

---

oxygen

67. and gives off ____________________.

---

carbon dioxide

1328. into the ____________ vein

dorsalis pedis

1329. which then empties into the ____________ vein.

anterior tibial

1330. From the anterior tibial vein the blood flows into the ____________ vein

popliteal

1331. and then into the ____________ vein.

femoral

1332. Another path for the blood to follow to the deep veins of the foot is from the venules to the plantar venous arch.

1333. From the venules of the foot the blood flows into the dorsal venous arch and the ____________ venous arch.

plantar

1334. The ____________ venous arch and the dorsal venous arch receive the blood flowing from the venules of the foot.

plantar

1335. The blood returning from the venules of the foot flows into the ____________ arch and the ____________ arch.

dorsal venous, plantar venous

1336. From the plantar venous arch the blood flows into the medial and lateral plantar veins.

1337. The veins which receive the blood flowing from the plantar venous arch are the medial and lateral ____________ veins.

plantar

1338. The blood flows from the plantar venous arch into the medial and ____________ plantar veins.

lateral

1339. The ____________ plantar veins receive the blood flowing from the plantar venous arch.

medial and lateral

1340. From the venules of the foot the blood flows into the ____________ arch and the ____________ arch.

dorsal venous, plantar venous

1341. From the plantar venous arch the blood flows into the ____________ veins.

medial and lateral plantar

68. The blood flows through the capillaries, around the alveoli, and then enters the pulmonary veins.

---

69. After the blood has picked up oxygen from the alveoli and at the same time given off carbon dioxide, it flows into the ________________ veins.

pulmonary

---

70. The pulmonary ___________ receive the blood after it passes around the alveoli.

veins

---

71. The blood flows to the lungs through the pulmonary artery. Then it flows into the capillaries, around the alveoli and then into the __________________.

pulmonary veins

---

72. The pulmonary veins then carry the blood away from the lungs.

---

73. The pulmonary _____________ carries the blood to the lungs.

artery

---

74. The pulmonary ___________ carry the blood from the lungs.

veins

---

75. Time for another review. (Unless you feel confident. If so, proceed to frame 86). Blood enters the heart at the ____________________,

right atrium

---

76. flows through the _______________ valve

tricuspid

---

77. into the ____________________________.

right ventricle

---

78. It then passes through the ________________ valves

semilunar

---

79. into the ____________________________

pulmonary artery

---

80. which carries the blood to the _______________.

lungs

---

81. In the lungs the pulmonary artery subdivides into __________________.

capillaries

---

82. These capillaries carry the blood around the tiny air sacs called __________________.

alveoli

1313. From the dorsalis pedis vein the blood flows into the anterior tibial vein.

1314. The anterior __________ vein receives the blood flowing from the dorsalis pedis vein.

tibial

1315. From the dorsalis pedis vein the blood flows into the __________ tibial vein.

anterior

1316. The blood returning from the foot through the deep veins flows from the venules into the dorsal venous arch into the dorsalis pedis vein and then into the __________ vein.

anterior tibial

1317. From the anterior tibial vein the blood flows into the popliteal vein.

1318. The vein which receives the blood flowing from the anterior tibial vein is the __________ vein.

popliteal

1319. From the anterior tibial vein the blood flows into the __________ vein.

popliteal

1320. The popliteal vein receives the blood flowing from the __________ vein.

anterior tibial

1321. As before, the popliteal vein empties into the __________ vein.

femoral

1322. Time for a review. (If you don't need one, go on to frame 1332).
The superficial path of the blood returning from the foot is through the venules to the __________.

dorsal venous arch

1323. From the dorsal venous arch the blood flows superficially into the __________ veins.

small and great saphenous

1324. The small saphenous vein empties into the __________ vein

popliteal

1325. which in turn empties into the __________ vein.

femoral

1326. The great saphenous vein empties into the __________ vein.

femoral

1327. The deep path of the blood flowing from the foot is from the venules into the __________

dorsal venous arch

---

83. From the alveoli the blood picks up ________________

oxygen

84. and gives up ________________________.

carbon dioxide

85. After leaving the alveoli the blood flows into the ____________________.

pulmonary veins

86. Label the indicate parts:

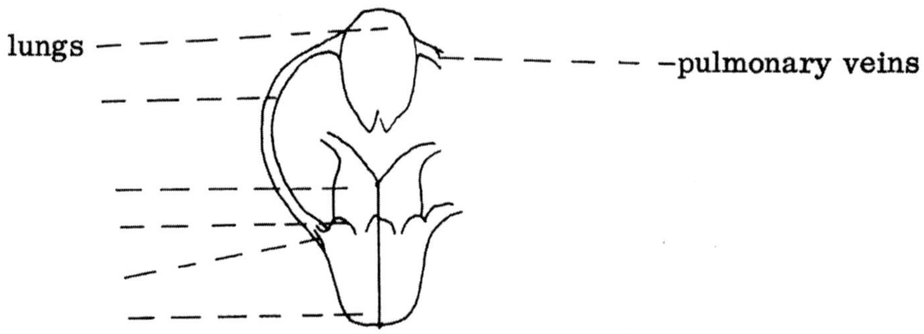

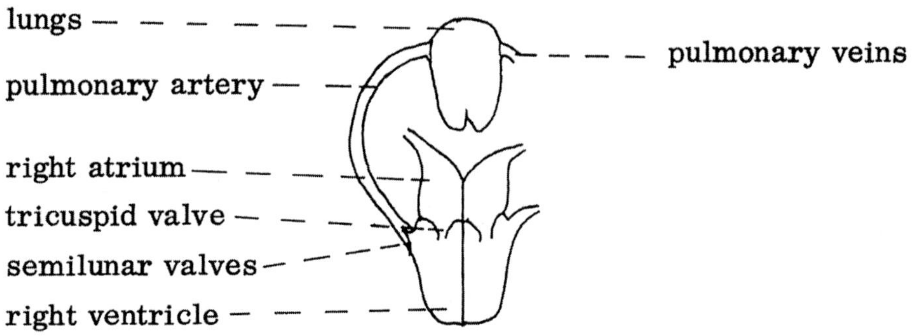

87. The pulmonary veins carry the blood to the left atrium of the heart.

88. The pulmonary veins carry the blood from the lungs to the left ______________.

atrium

89. From the lungs the blood flows into the _____________ atrium.

left

90. The veins which carry the blood from the lungs to the left atrium are called the _____________ veins.

pulmonary

91. The blood flows to the lungs through the ________________ artery.

pulmonary

92. The blood flows from the lungs through the ________________ veins.

pulmonary

1298. From the popliteal vein the blood flows into the (deep) femoral vein.

1299. The ______________ vein receives the blood flowing from the popliteal vein.

femoral

1300. From the popliteal vein the blood flows into the ______________ vein.

femoral

1301. The great saphenous vein empties directly into the femoral vein.

1302. The ______________ vein receives the blood flowing from the great saphenous vein.

femoral

1303. From the great saphenous vein the blood flows into the ______________ vein.

femoral

1304. The blood returning from the foot flows into the venules and then the ______________.

dorsal venous arch

1305. From the dorsal venous arch the superficial path of the blood is into the ______________ veins.

small and great saphenous

1306. From the small saphenous vein the blood flows into the ______________ vein

popliteal

1307. and then into the ______________ vein.

femoral

1308. From the great saphenous vein the blood flows into the ______________ vein.

femoral

1309. From the dorsal venous arch there is also a deep set of veins returning the blood to the heart. The blood can flow from the dorsal venous arch into the dorsalis pedis vein.

1310. The vein which receives the blood flowing to the deep veins from the dorsal venous arch is the ______________ pedis vein.

dorsalis

1311. From the dorsal venous arch the blood flows to the deep veins through the dorsalis ______________ vein.

pedis

1312. The blood returns from the foot through the venules, through the dorsal venous arch and then, for the deep veins, into the ______________ vein.

dorsalis pedis

93. The pulmonary veins carry the blood to the ______________.

left atrium

94. From the left atrium the blood passes through the <u>bicuspid valve</u>.

95. The valve by which the blood leaves the left atrium is the ______________ valve.

bicuspid

96. The blood enters the left atrium from the pulmonary veins and leaves through the ______________.

bicuspid valve

97. In the lungs the blood picks up ______________ from the alveoli,

oxygen

98. and at the same time gives off ______________ to the alveoli.

carbon dioxide

99. From the alveoli the blood flows into the ______________ veins.

pulmonary

100. which in turn carry the blood to the ______________ of the heart.

left atrium

101. From the left atrium the blood passes through the ______________ valve.

bicuspid

102. Label the following diagram:

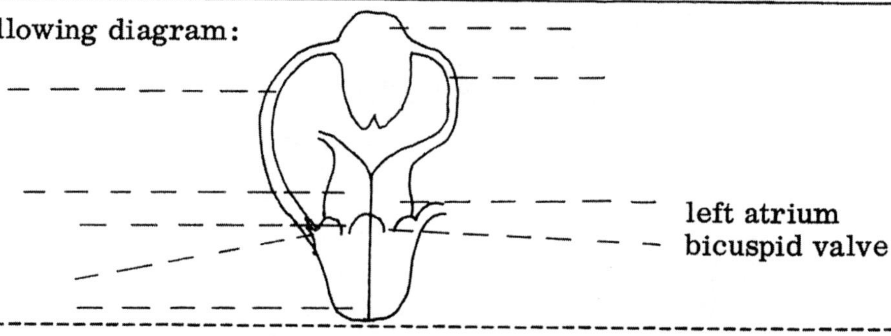

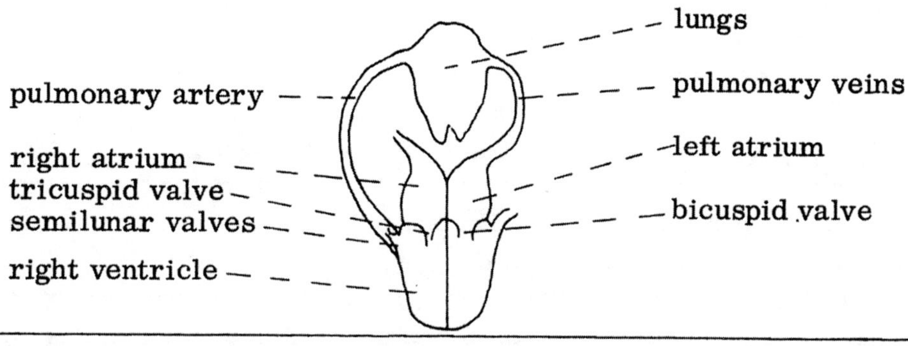

103. From the lungs the blood flows through the pulmonary veins to the left atrium. Then it flows through the bicuspid valve into the <u>left ventricle</u>.

1285. The blood flows from the venules into the dorsal venous arch and then, superficially, into the small saphenous vein (along the back of the leg) or the great saphenous vein (along the medial side of the leg).

1286. From the dorsal venous arch the superficial path of the blood is into the small saphenous vein or the great ________________ vein.

saphenous

1287. The blood flows from the dorsal venous arch along the back of the leg through the small ________________ vein.

saphenous

1288. The blood flows from the dorsal venous arch along the medial side of the leg through the ________________ saphenous vein.

great

1289. From the dorsal venous arch the blood flows superficially through the small or great ________________ veins.

saphenous

1290. The vein which carries the blood superficially from the dorsal venous arch along the back of the leg is the ________________ vein.

small saphenous

1291. The vein which carries the blood superficially from the dorsal venous arch along the medial side of the leg is the ________________ vein.

great saphenous

1292. The blood returning from the foot flows into the venules and then into the ________________.

dorsal venous arch

1293. From the dorsal venous arch the superficial path of the blood is through the ________________ veins.

small and great saphenous

1294. The small saphenous vein empties into the (deep) popliteal vein.

1295. The vein which receives the blood flowing from the small saphenous vein is the ________________ vein.

popliteal

1296. The blood returning along the back of the leg flows through the ________________ vein

small saphenous

1297. and then into the ________________ vein.

popliteal

104. The blood flows from the left atrium to the ______________ ventricle through the bicuspid valve.

---

left

105. The left ______________ receives the blood flowing through the bicuspid valve.

---

ventricle

106. The blood flows from the left atrium through the bicuspid valve to the ______________.

---

left ventricle

107. Time for one more review. (But if you are sure of the material, go on to frame 121). The blood enters the heart at the ______________,

---

right atrium

108. flows through the ______________

---

tricuspid valve

109. into the ______________.

---

right ventricle

110. From the right ventricle the blood flows through the ______________ valves

---

semilunar

111. into the ______________

---

pulmonary artery

112. which carries the blood to the ______________.

---

lungs

113. In the lungs the pulmonary artery subdivides into many branches called ______________.

---

capillaries

114. These capillaries carry the blood around the tiny air sacs called ______________.

---

alveoli

115. From the alveoli the blood picks up ______________

---

oxygen

116. and gives off ______________.

---

carbon dioxide

117. The blood then flows into the ______________ veins

---

pulmonary

118. which carry the blood to the ______________ of the heart.

---

left atrium

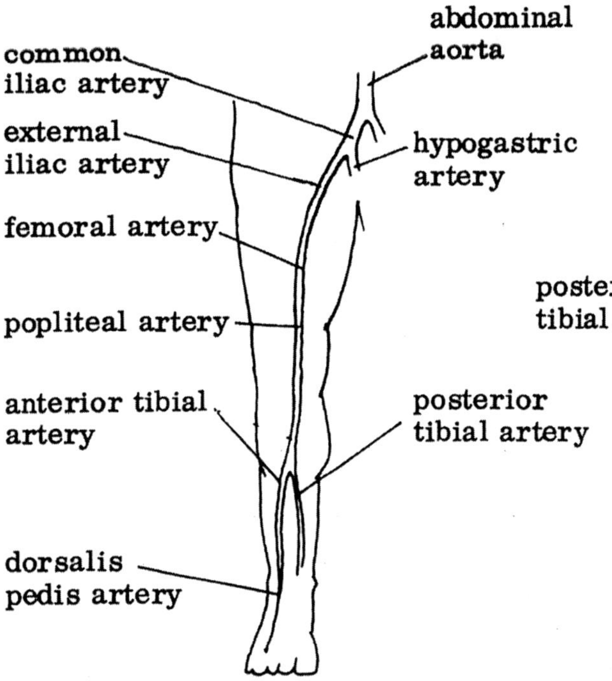

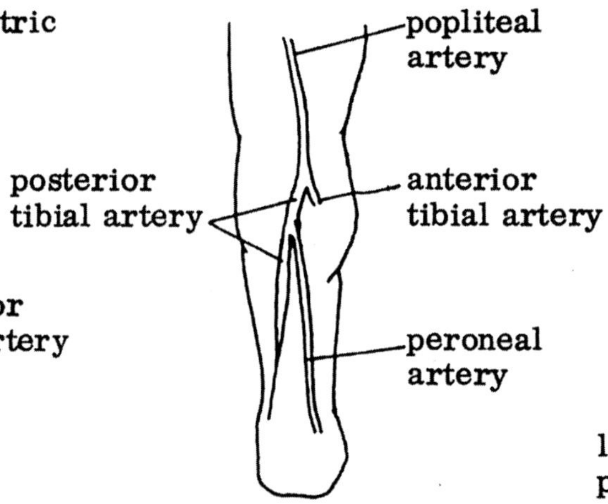

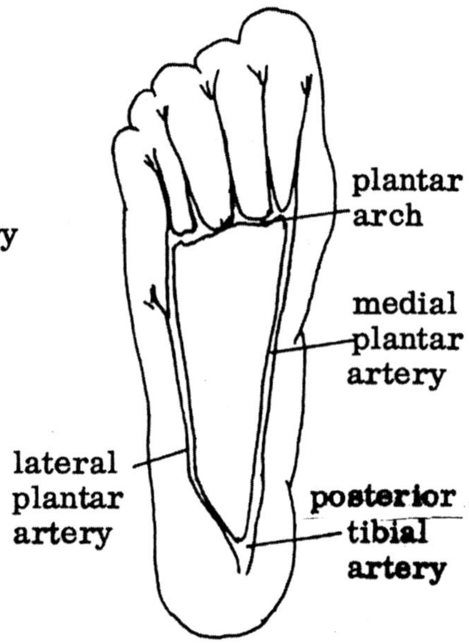

1278. **From the plantar arch the blood flows to all parts of the foot through small arteries,** arterioles, and finally capillaries.

1279. **Trace the path of the blood from the heart to the foot.**

Heart to ascending aorta to arch of aorta to thoracic aorta to abdominal aorta to right or left common iliac artery to right or left external iliac artery to femoral artery to popliteal artery to anterior and posterior tibial arteries. From anterior tibial artery to dorsalis pedis artery to plantar arch of the foot. From posterior tibial artery to peroneal artery and also to medial and lateral plantar arteries to plantar arch. From plantar arch to arterioles and capillaries of the foot.

1280. The blood returning from the foot flows into the venules of the foot and then the dorsal venous arch.

1281. The venous blood flows from the venules into the dorsal ______________ arch.

---

venous

1282. The ____________ venous arch receives the blood flowing from the venules of the foot.

---

dorsal

1283. The blood flows from the plantar arch to the arterioles, to the capillaries, to the venules and then into the _______________ arch.

---

dorsal venous

1284. From the dorsal venous arch the blood flows either through a superficial set of veins or a deep set of veins.

---

119. From the left atrium the blood flows through the ____________________

---

bicuspid valve

120. into the ________________.

---

left ventricle

121. From the left ventricle the blood passes through another set of <u>semilunar</u> <u>valves</u>.

122. When the blood leaves the left ventricle, it passes through the ______________ valves.

---

semilunar

123. The ________________ valves receive the blood flowing from the left ventricle.

---

semilunar

124. From the left atrium the blood flows through the ______________ valve

---

bicuspid

125. into the ________________________.

---

left ventricle

126. From the left ventricle the blood flows through the _________________ valves.

---

semilunar

127. From the semilunar valves the blood flows into the main artery, the <u>aorta</u> ($\bar{a}$ or' ta).

128. The artery which receives the blood from the left ventricle (after passing through the semilunar valves) is called the ______________.

---

aorta

129. Blood flows from the lungs through the _______________veins,

---

pulmonary

130. and enters the heart at the ___________________.

---

left atrium

131. From the left atrium the blood flows through the ______________________

---

bicuspid valve

132. into the ______________________.

---

left ventricle

133. From the left ventricle the blood flows through the ____________________ valves

---

semilunar

134. into the _______________.

---

aorta

1272. The anterior tibial artery extends along the anterior side of the leg to the ankle where it becomes the ____________ artery

dorsalis pedis

1273. leading to the ______________ of the foot.

plantar arch

1274. The posterior tibial artery extends along the back of the leg, giving off below the knee a branch called the ____________ artery.

peroneal

1275. At the ankle the posterior tibial artery divides into the ________________ arteries

medial and lateral plantar

1276. which combine to form the _______________ in the foot.

plantar arch

1277. Label

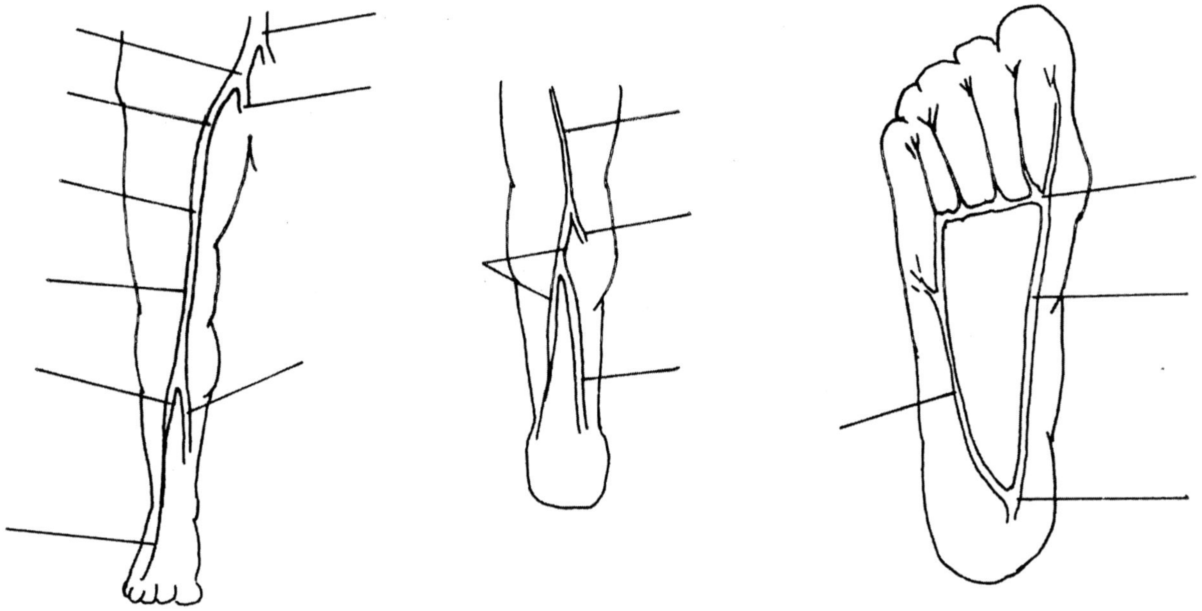

135. Label:

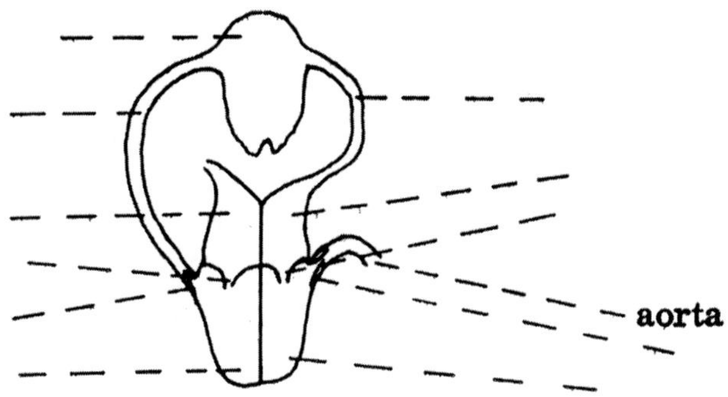

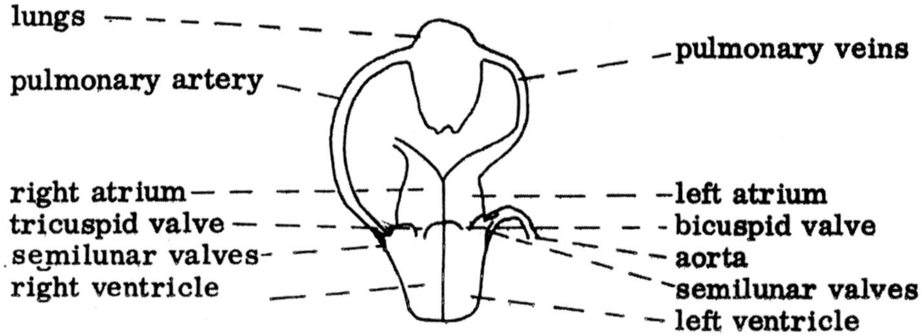

---

136. From the aorta the blood goes to all parts of the body, eventually flowing into the vena cava from which it enters the right atrium to begin the cycle again.

---

137. The artery by which the blood leaves the left ventricle is called the ____________.

aorta

---

138. From the aorta the blood travels to all parts of the body, finally returning to the ________ cava.

vena

---

139. The vena __________ receives the blood returning from all parts of the body.

cava

---

140. After the blood flows through the aorta to all parts of the body, it flows into the ____________________.

vena cava

---

141. From the vena cava the blood flows into the ____________ atrium to begin a new cycle.

right

---

142. Time for a final review. (Last chance to review, but if you are sure of yourself, go on to frame 160).
Blood enters the heart at the ______________________,

right atrium

1262. The blood flows from the medial and lateral plantar arteries into the __________ arch.

plantar

1263. The plantar ______________ receives the blood flowing from the medial and plantar arteries.

arch

1264. The blood from the anterior tibial artery flows into the ____________ artery

dorsalis pedis

1265. and then into the _____________.

plantar arch

1266. The blood flows from the medial and lateral plantar arteries into the _____________.

plantar arch

1267. Label:

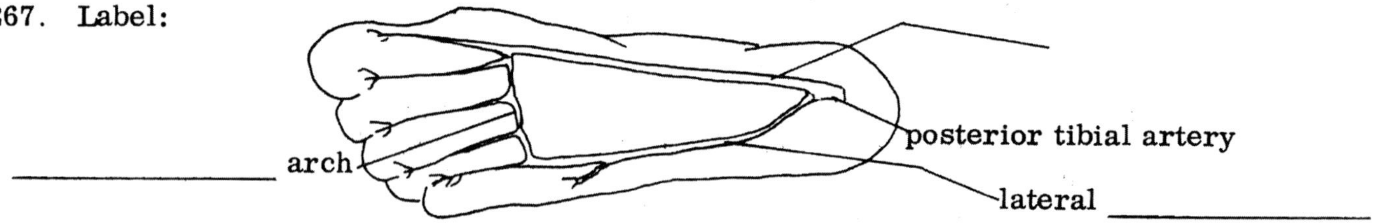

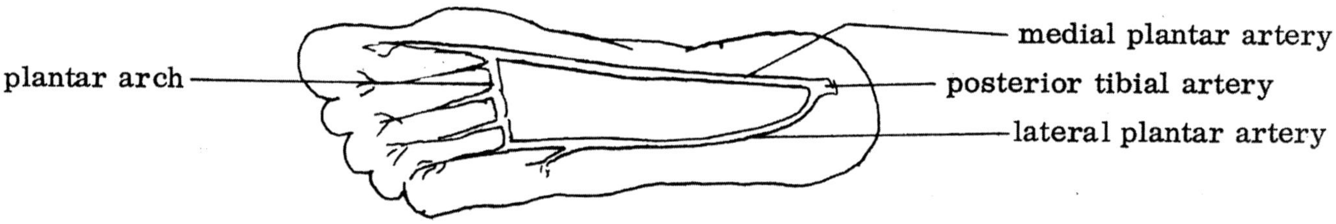

1268. Time for a review: (If you don't need one, go on to frame 1277).
The blood flows from the common iliac artery into the __________ artery and the __________ artery.

external iliac, hypogastric

1269. From the external iliac artery the blood flows into the ______________ artery

femoral

1270. and then into the ______________ artery.

popliteal

1271. The popliteal artery subdivides into the ____________ artery and the ____________ artery.

anterior tibial, posterior tibial

143. flows through the ___________________

tricuspid valve

144. into the ______________________.

right ventricle

145. The blood leaves the right ventricle through the __________________ valves

semilunar

146. and flows through the ______________________________

pulmonary artery

147. to the _________________.

lungs

148. In the lungs the pulmonary artery subdivides into __________________.

capillaries

149. These capillaries carry the blood around the air sacs, the _______________.

alveoli

150. From the alveoli the blood picks up ________________

oxygen

151. and gives off ___________________.

carbon dioxide

152. The blood then flows into the ___________________veins

pulmonary

153. and enters the heart at the ______________________.

left atrium

154. From the left atrium the blood passes through the ______________________

bicuspid valve

155. into the _______________________.

left ventricle

156. From the left ventricle the blood flows through the ____________________ valves

semilunar

157. into the ___________________,

aorta

158. which carries the blood to all parts of the body. The blood then flows into the _____________.

vena cava

1252. At the ankle the posterior tibial artery divides into the medial and lateral plantar arteries.

1253. The posterior tibial artery subdivides at the ankle into the medial and lateral ______________ arteries.

plantar

1254. The two divisions of the posterior tibial artery are the medial and the ______________ plantar arteries.

lateral

1255. The posterior tibial artery divides into the ______________ and the lateral plantar arteries at the ankle.

medial

1256. The ______________ plantar arteries are the branches of the posterior tibial artery at the ankle.

medial and lateral

1257. The blood flows from the posterior tibial artery to the foot through the __________ arteries.

medial and lateral plantar

1258. The popliteal artery divides into the __________ artery and the __________ artery.

anterior tibial, posterior tibial

1259. The posterior tibial artery gives off one large branch just below the knee, the ________________ artery,

peroneal

1260. and then at the ankle divides into the ______________________ arteries.

medial and lateral plantar

1261. The medial and lateral plantar arteries form the plantar arch which also receives blood from the anterior tibial artery.

159. The vena cava empties into the _________________ to begin a new cycle.

right atrium

160. Label:

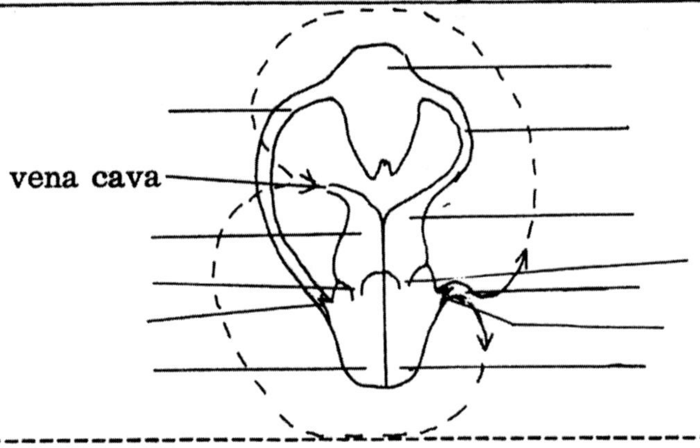

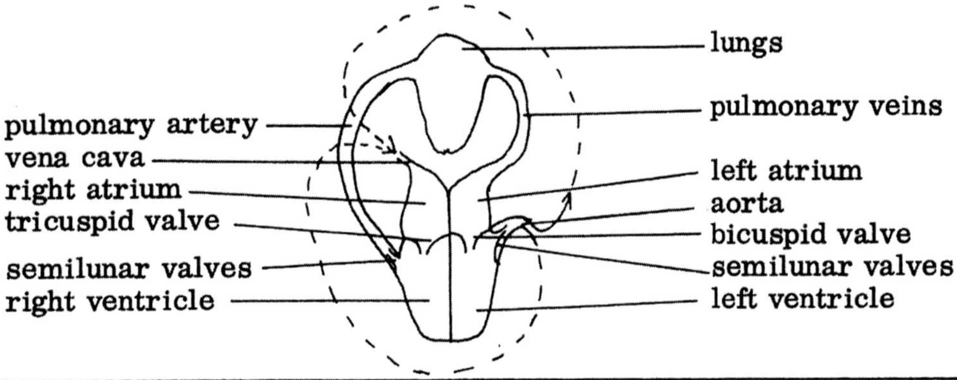

161. Describe the circulation of the blood through the heart, beginning with the right atrium.

Right atrium, through tricuspid valve to right ventricle, through semilunar valves to pulmonary artery to lungs to capillaries, around alveoli to pick up oxygen and give off carbon dioxide, to pulmonary veins to left atrium through bicuspid valve to left ventricle through semilunar valves to aorta to all parts of the body to vena cava and return to right atrium.

1240. The ______________ receives the blood flowing from the dorsalis pedis artery.

plantar arch

1241. The blood flows from the anterior tibial artery into the ____________ artery

dorsalis pedis

1242. and from there into the ______________ of the foot.

plantar arch

1243. The popliteal artery subdivides into the __________ artery and the __________ artery.

anterior tibial, posterior tibial

1244. The posterior tibial artery extends along the back of the leg to the ankle. Just below the knee it gives off one large branch, the peroneal artery.

1245. The peroneal artery is a branch of the posterior ________________ artery.

tibial

1246. The posterior tibial artery extends from the popliteal artery to the ankle. It gives off one large branch, the ______________ artery.

peroneal

1247. The ________________ artery is a branch of the posterior tibial artery.

peroneal

1248. The femoral artery extends into the ________________ artery.

popliteal

1249. The popliteal artery divides into the ______________ artery and the ______________ artery.

anterior tibial, posterior tibial

1250. The posterior tibial artery extends along the back of the leg and gives off one large branch, the ________________ artery.

peroneal

1251. Label the indicated arteries of the back of the leg:

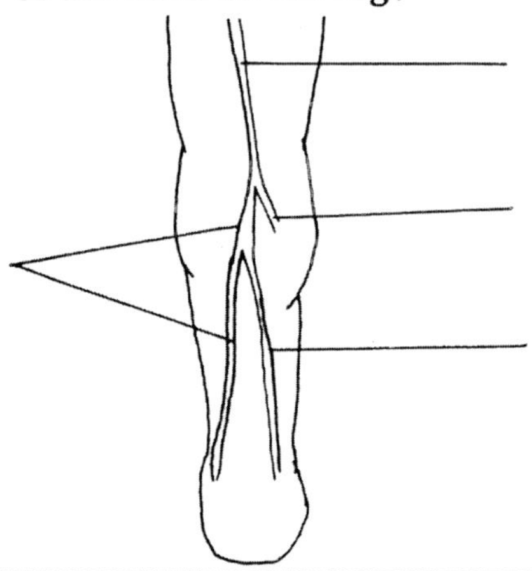

II. THE ASCENDING AORTA - CORONARY CIRCULATION

162. The aorta arises from the left ventricle and arches backward, descending to the left. For descriptive purposes it may be considered as being subdivided into:

the ascending aorta
the arch of the aorta
the thoracic aorta
the abdominal aorta

---

163. This part of the program will deal only with the ascending aorta which is indicated in the following diagram.

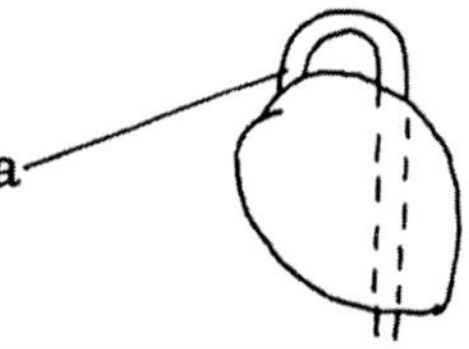

---

164. The ascending aorta has two small branches, the right and left coronary arteries which supply blood to the heart muscle.

---

165. The ascending aorta contains two branches, the ________ and ________ coronary arteries.

right, left

---

166. The arteries which supply blood to the heart muscle are called the right and left ________ arteries.

coronary

---

167. The coronary arteries arise from the ____________ aorta.

ascending

---

168. The heart muscle receives its blood supply from the ____________ arteries which branch-off the ____________ aorta.

coronary, ascending

---

169. The coronary arteries divide into many tiny branches called capillaries.

---

170. The many tiny branches of the right and left coronary arteries are called ____________.

capillaries

---

171. The heart receives its nourishment from the blood as it flows through the tiny ________.

capillaries

---

172. From the capillaries the hearts blood supply flows into the cardiac veins.

---

173. The veins which receive the blood flowing from the capillaries of the heart are called the ____________ veins.

cardiac

1235. The continuation of the anterior tibial artery is the ______________ artery.

dorsalis pedis

1236. Label:

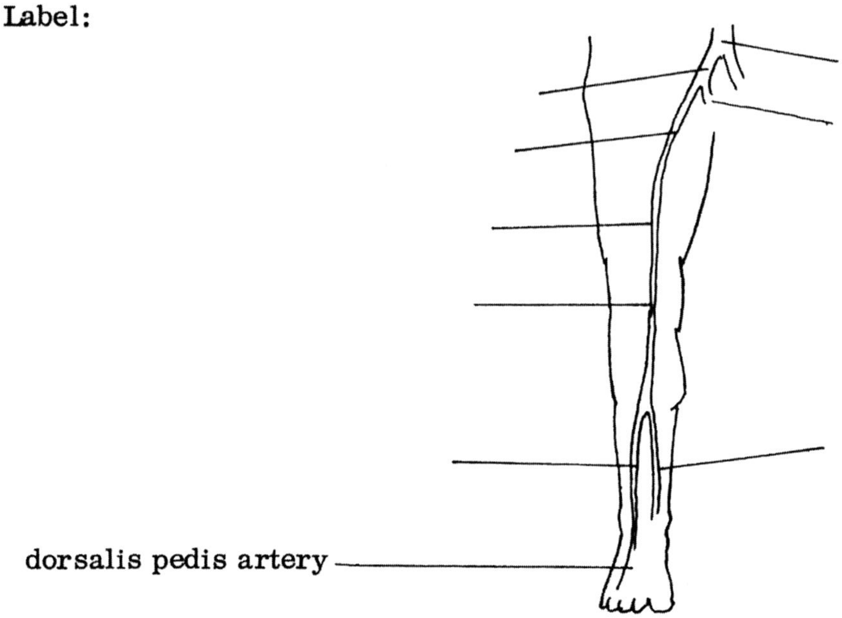

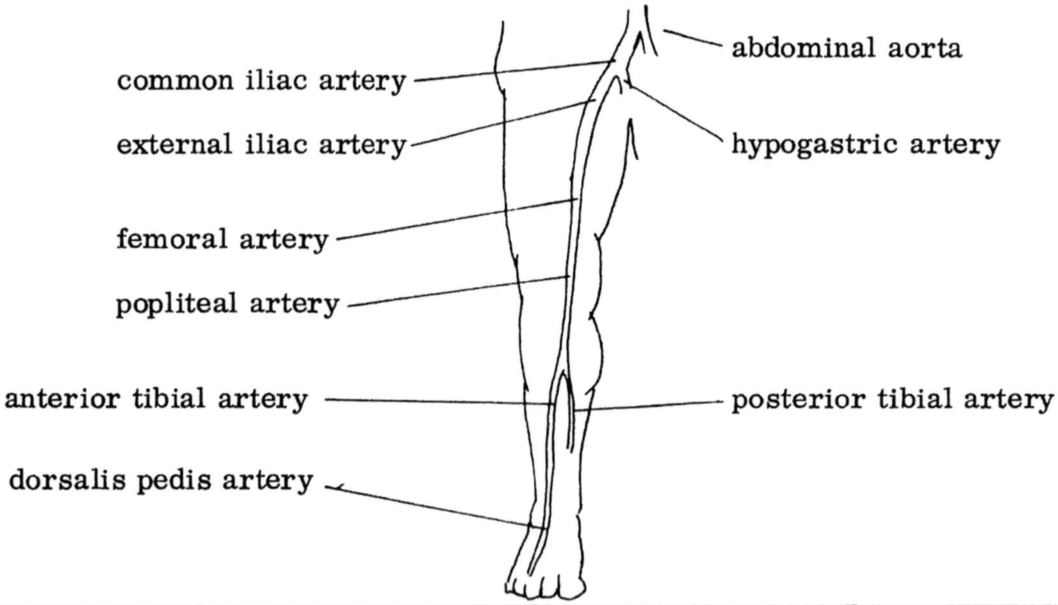

1237. From the dorsalis pedis artery the blood flows to the plantar arch of the foot.

1238. The blood flows from the dorsalis pedis artery to the ____________ arch.

plantar

1239. From the dorsalis pedis artery the blood flows into the plantar ___________ of the foot.

arch

174. The ____________ veins receive the blood flowing from the capillaries of the heart.

---

cardiac

175. Blood flows to the heart muscle through the right and left coronary arteries to the capillaries and then to the ____________ ____________.

---

cardiac veins

176. Reviewing (if you feel confident and don't need a review, proceed to frame 180).
Blood leaving the heart enters the ____________ aorta.

---

ascending

177. The ascending aorta has two branches, the __________ and the ____________ ____________ arteries.

---

right, left, coronary

178. The right and left coronary arteries subdivide into many ____________.

---

capillaries

179. These capillaries empty into the ______________________.

---

cardiac veins

180. From the cardiac veins the heart's blood supply flows into the <u>coronary sinus</u>.

181. Blood flows from the heart capillaries into the cardiac veins and then empties into the ________________ sinus.

---

coronary

182. The ____________ sinus is the vein which receives the blood after it has nourished the heart.

---

coronary

183. The blood flows through the right and left coronary arteries, through the heart capillaries, through the cardiac veins and then into the ___________________.

---

coronary sinus

184. From the coronary sinus the blood enters the <u>right</u> <u>atrium</u>.

185. The coronary sinus empties into the right ____________.

---

atrium

186. The __________ atrium receives the blood flowing from the coronary sinus.

---

right

187. The heart's blood supply flows from the coronary sinus into the ______________________.

---

right atrium

188. Review time again. (If you feel that you know the coronary circulation, proceed to frame 195).
Blood flows from the left ventricle into the _____________ aorta.

---

ascending

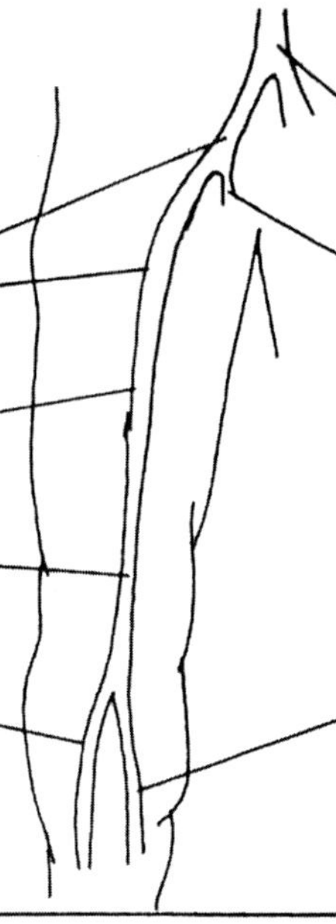

1225. The anterior tibial artery continues along the anterior part of the leg to the ankle joint where it becomes the dorsalis pedis artery.

1226. When the anterior tibial artery reaches the ankle joint it becomes the dorsalis ________ artery.

pedis

1227. The continuation of the anterior tibial artery is the dorsalis ________ artery.

pedis

1228. The anterior tibial artery continues along the anterior part of the leg until it reaches the ankle joint where it becomes the ________ pedis artery.

dorsalis

1229. The ________ pedis artery is a continuation of the anterior tibial artery.

dorsalis

1230. At the ankle joint the blood flows from the anterior tibial artery into the ________ artery.

dorsalis pedis

1231. The artery extending from the ankle joint to the foot is the ________ artery.

dorsalis pedis

1232. The continuation of the external iliac artery is the ________ artery.

femoral

1233. The continuation of the femoral artery is the ________ artery.

popliteal

1234. The continuation of the popliteal artery is the ________ artery and the ________ artery.

anterior tibial, posterior tibial

189. The ascending aorta contains two branches, the right and left ________________

coronary arteries

190. which supply blood for the nourishment of the ____________.

heart

191. The blood from the coronary arteries flows through many branches called ____________.

capillaries

192. From the capillaries the blood flows into veins called ________________.

cardiac veins

193. From the cardiac veins the blood then enters the ________________

coronary sinus

194. which empties into the ____________.

right atrium

195. Trace the path of the blood as it flows to nourish the heart.

From heart to ascending aorta to right and left coronary arteries to heart capillaries to cardiac veins to coronary sinus to right atrium.

III. THE ARCH OF THE AORTA

196. The purpose of this part of the program is to trace the circulation of the blood as it flows through the arch of the aorta.

197. The arch of the aorta contains three branches. The first and largest branch is called the innominate artery. (i-nom'-i-nit)

198. The first branch of the arch of the aorta is called the in _ om _ _ ate artery.

innominate

199. The ______________ artery is the largest branch of the aortic arch.

innominate

200. As the blood flows through the aortic arch, it first reaches the ____________ artery.

innominate

1214. The extension of the external iliac artery to the upper part of the thigh is called the ________________ artery.

femoral

1215. The extension of the femoral artery is known as the ________________ artery.

popliteal

1216. Just below the knee the popliteal artery subdivides into the anterior tibial artery and the posterior tibial artery.

1217. The popliteal artery subdivides into the posterior tibial artery and the ________________ tibial artery.

anterior

1218. The two branches of the popliteal artery are the posterior tibial artery and the ________________ artery.

anterior tibial

1219. The popliteal artery divides into the anterior ______________ artery and the posterior ______________ artery.

tibial, tibial

1220. Just below the knee the popliteal artery subdivides into the ________________ artery and the ________________ artery.

anterior tibial, posterior tibial

1221. The ________________ artery is a continuation of the external iliac artery.

femoral

1222. The ________________ artery is a continuation of the femoral artery.

popliteal

1223. The ____________ artery and the ____________ artery continue from the popliteal artery.

anterior tibial, posterior tibial

1224. Label:

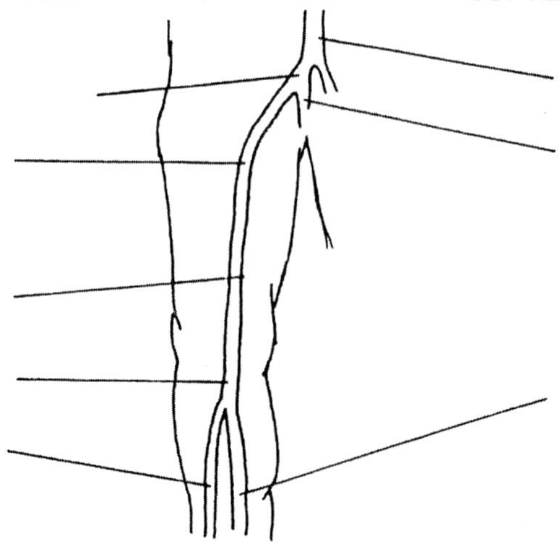

posterior ________________

201. Label the following diagram:

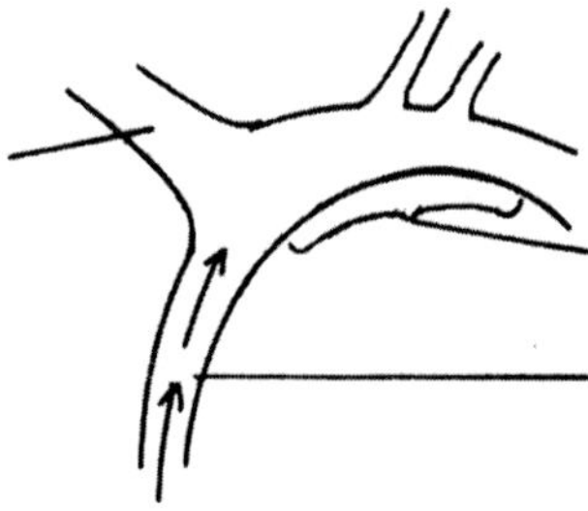

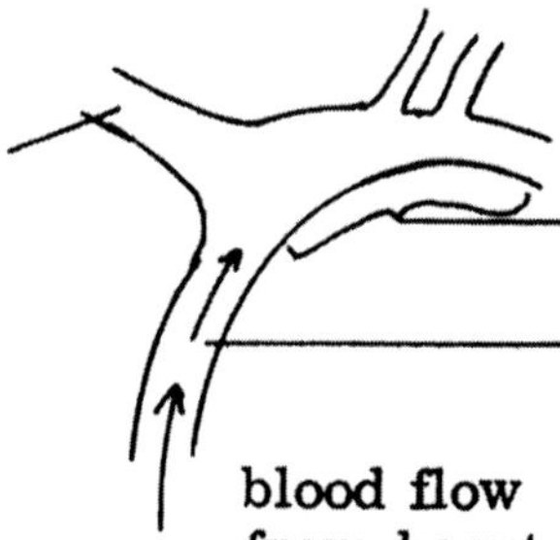

---

202. The second branch of the aortic arch is called the <u>left common carotid artery</u>. (ka-rot'-id)

---

203. The branch of the aortic arch after the innominate artery is called the left common ______________ artery.

carotid

---

204. The second branch of the aortic arch is called the ____________ common ____________ artery.

left, carotid

---

205. The __________ __________ __________ artery is the second branch of the aortic arch.

left common carotid

---

206. The first branch of the aortic arch is called the _______________ artery.

innominate

---

207. The second branch of the aortic arch is called the ______________________________ artery.

left common carotid

1204. The ________________ artery is the extension of the external iliac artery.

femoral

1205. At its terminal end the abdominal aorta divides into the right and left ____________ arteries.

common iliac

1206. The common iliac arteries subdivide into the ____________ arteries and the ____________ arteries.

external iliac, hypogastric

1207. The external iliac artery continues through the upper three-fourths of the thigh where it is called the _______________ artery.

femoral

1208. Label:

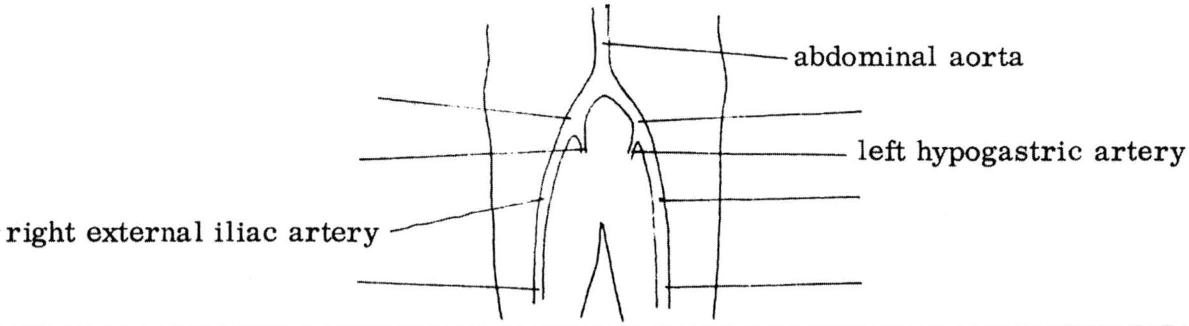

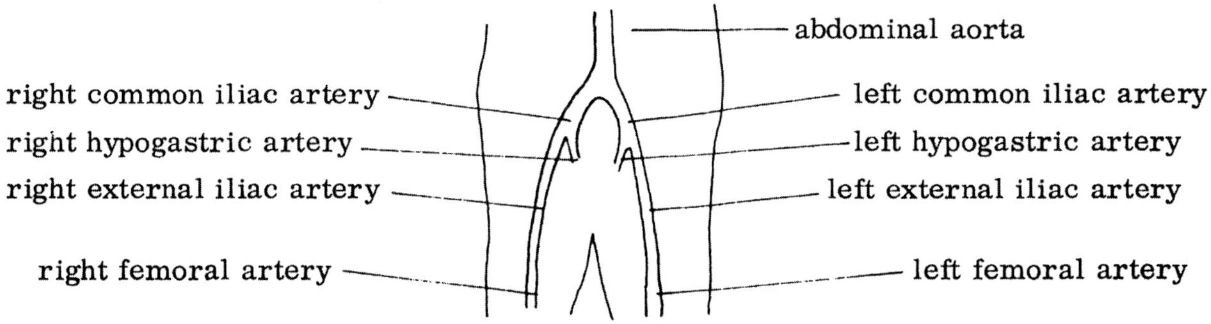

1209. At the back of the knee the femoral artery becomes the popliteal artery.

1210. The continuation of the femoral artery is the _____________ artery.

popliteal

1211. The ________________ artery is a continuation of the femoral artery.

popliteal

1212. The ______________ artery is a continuation of the external iliac artery.

femoral

1213. The common iliac artery divides into the __________ artery and the __________ artery.

external iliac, hypogastric

208. Label the following diagram:

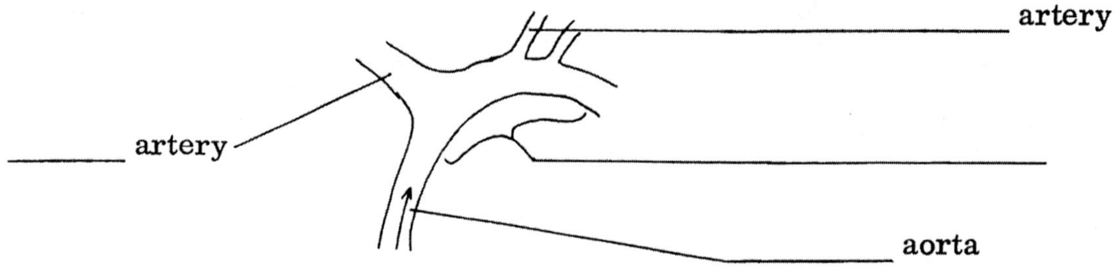

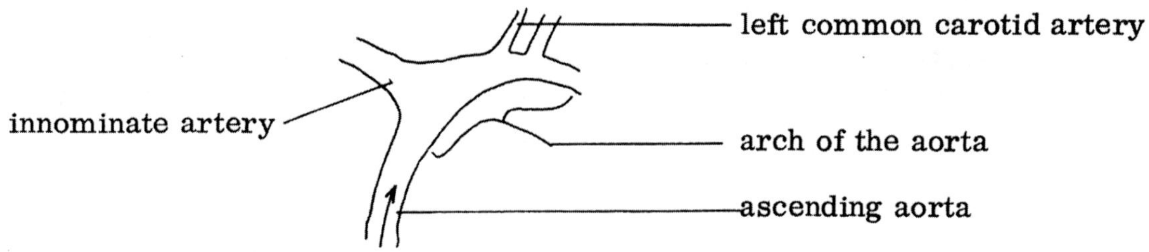

---

209. The third branch of the aortic arch is called the <u>left subclavian artery</u>. (sub-klā'-vi-en)

---

210. The last branch of the arch of the aorta is called the left sub__________ artery.

subclavian

---

211. After the left common carotid artery, the next branch in the aortic arch is the left _____________ artery.

subclavian

---

212. The ___________________ artery is the third branch of the aortic arch.

left subclavian

---

213. The first branch of the aortic arch is the ___________ artery.

innominate

---

214. The second branch of the aortic arch is the _______________________ artery.

left common carotid

---

215. The third branch of the aortic arch is the ____________________ artery.

left subclavian

1195. Trace the path of the blood from the heart to the under surface of the diaphragm and return.

From heart to ascending aorta to arch of aorta to thoracic aorta to abdominal aorta to inferior phrenic arteries to under surface of diaphragm to inferior phrenic veins to inferior vena cava to heart.

1196. Trace the path of the blood from the heart to the intestines and return.

From heart to ascending aorta to arch of aorta to thoracic aorta to abdominal aorta to superior mesenteric artery and to inferior mesenteric artery. From superior mesenteric artery to small intestine and half of large intestine to superior mesenteric vein to portal vein to liver to hepatic veins to inferior vena cava to heart. From inferior mesenteric vein to the other half of large intestine to inferior mesenteric vein to splenic vein to portal vein to liver to hepatic veins to inferior vena cava to heart.

1197. Trace the path of the blood from the heart to the suprarenal glands and return.

From heart to ascending aorta to arch of aorta to thoracic aorta to abdominal aorta to right and left suprarenal arteries to suprarenal glands to right and left suprarenal veins. From left suprarenal vein to left renal vein to inferior vena cava to heart. From right suprarenal vein to inferior vena cava to heart.

1198. Trace the path of the blood from the heart to the sacrum and coccyx and return.

From heart to ascending aorta to arch of aorta to thoracic aorta to abdominal aorta to middle sacral artery to sacrum and coccyx to middle sacral vein to inferior vena cava to heart.

1199. Trace the path of the blood from the heart to the ovaries and return.

From heart to ascending aorta to arch of aorta to thoracic aorta to abdominal aorta to right and left ovarian arteries to ovaries to right and left ovarian veins. From left ovarian vein to left renal vein to inferior vena cava to heart and from right ovarian vein to inferior vena cava to heart.

IX. THE LOWER EXTREMITIES

1200. At its terminal end the abdominal aorta divides into the right and left common iliac arteries Each of these subdivides into the external iliac artery and the hypogastric artery. The external iliac arteries extend to the lower extremities.

1201. The right external iliac artery extends to the right leg and the ______________ artery to the left leg.

left external iliac

1202. The external iliac artery extends towards the thigh where it is called the femoral artery.

1203. The artery extending through the upper three-fourths of the thigh is the ____________ artery.

femoral

216. Label the following diagram:

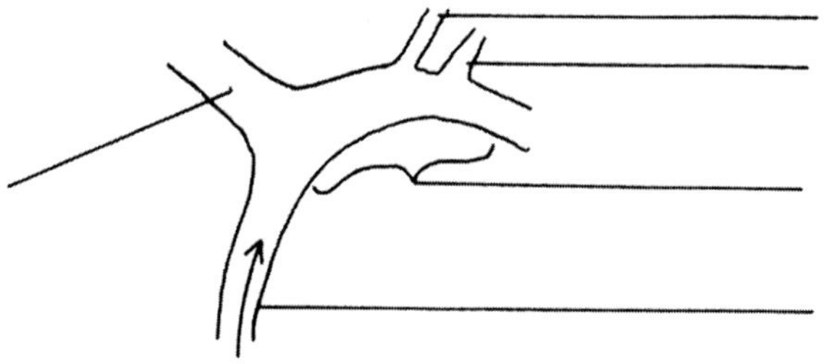

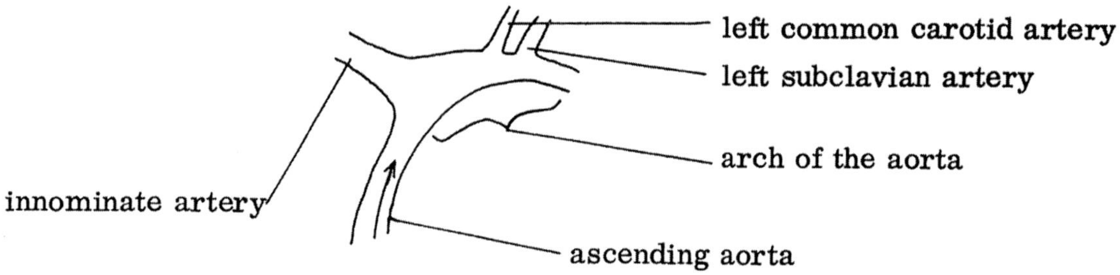

---

217. The first branch of the aortic arch is the ______________________.

innominate artery

---

218. The innominate artery divides into two branches. The branch going upward is called the right common carotid artery.

---

219. The upward branch of the innominate artery is the right __________ __________ artery.

common carotid

---

220. The first branch of the aortic arch is called the ___________ artery.

innominate

---

221. The first branch of the innominate artery is called the _________ common __________ artery.

right, carotid

---

222. The upward branch of the innominate artery is called the ____________________ artery.

right common carotid

---

223. The right common carotid artery is the first branch of the _________________ artery.

innominate

---

224. The other branch of the innominate artery extends towards the right arm. It is called the right subclavian artery.

---

225. The second branch of the innominate artery is called the right ___________ artery.

subclavian

1192. Label the indicated parts of the abdominal aorta:

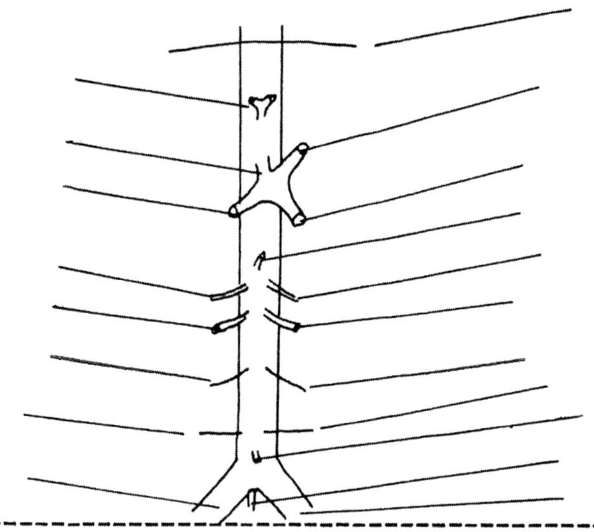

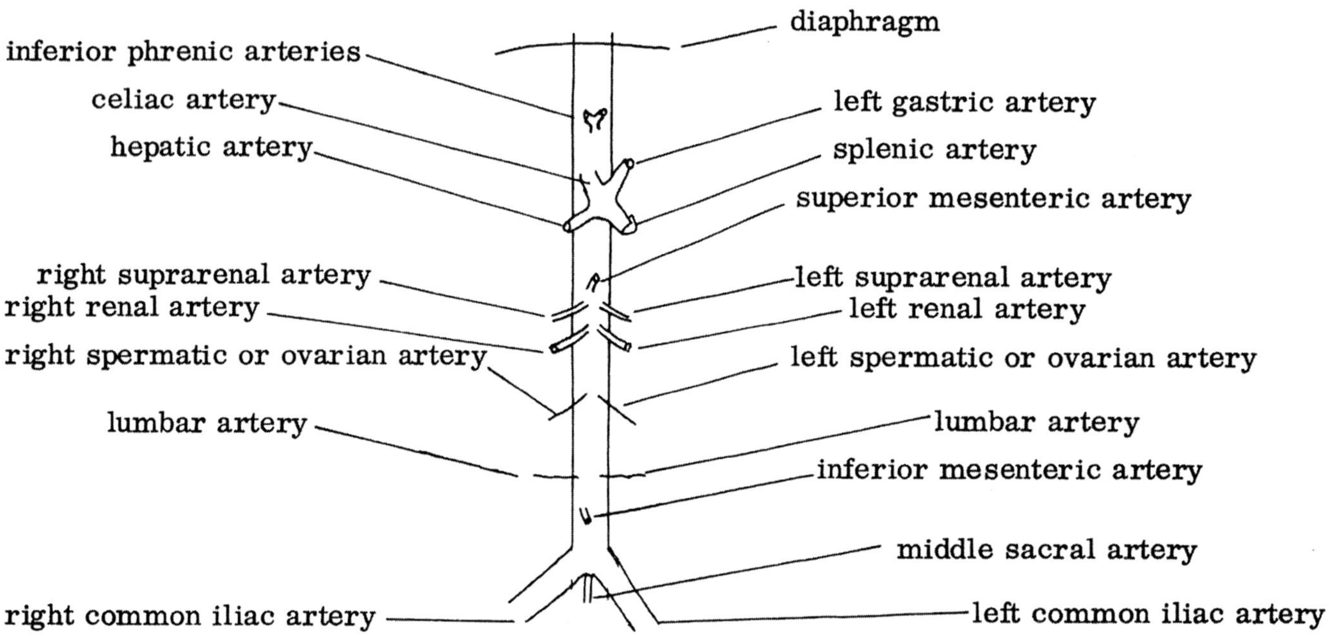

1193. Trace the path of the blood from the heart to the kidneys and return.

From heart to ascending aorta to arch of aorta to thoracic aorta to abdominal aorta to right and left renal arteries to kidneys to right and left renal veins to inferior vena cava and back to heart.

1194. Trace the path of the blood from the heart to the pelvis and return.

From heart to ascending aorta to arch of aorta to thoracic aorta to abdominal aorta to right and left common iliac arteries to hypogastric arteries to pelvis to hypogastric veins to common iliac veins to inferior vena cava to heart.

226. The __________ subclavian artery is the branch of the innominate artery that extends towards the right arm.

---

right

---

227. The innominate artery has two branches, the right common carotid artery and the __________________ artery.

---

right subclavian

---

228. The first branch of the aortic arch is the ______________ artery.

---

innominate

---

229. The first branch of the innominate artery is the ____________________ artery.

---

right common carotid

---

230. The second branch of the innominate artery is called the _________________ artery.

---

right subclavian

---

231. Label the following diagram:

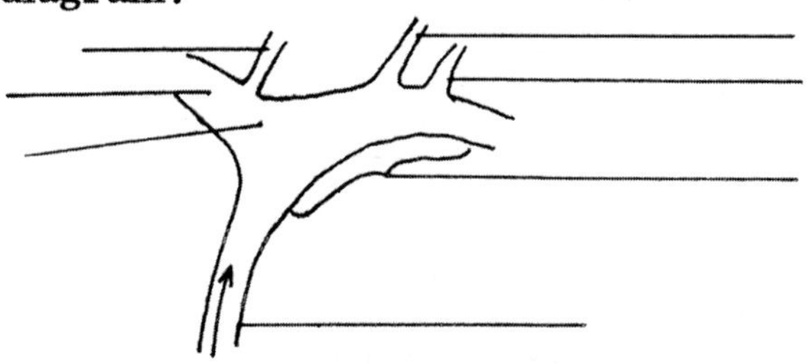

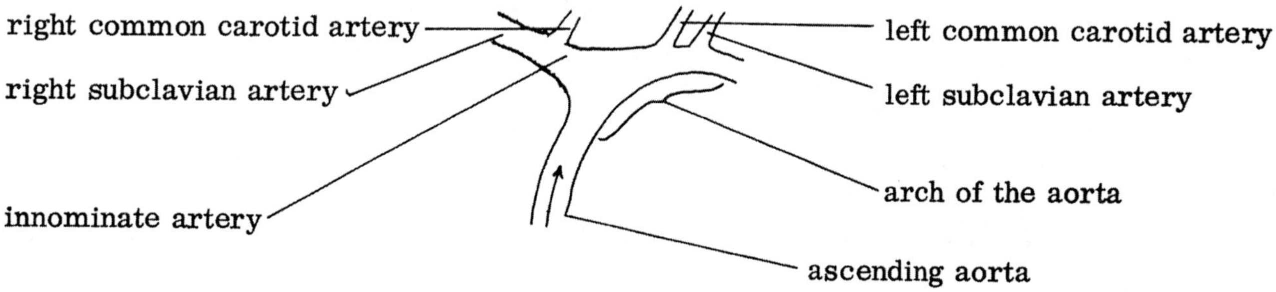

---

232. Branching off both the right and left subclavian arteries and extending upward towards the head are the vertebral arteries - the right vertebral artery from the right subclavian artery and the left vertebral artery from the left subclavian artery.

---

233. The vertebral arteries extend as branches of the right and left ____________ arteries.

---

subclavian

---

234. One branch of the right subclavian artery, extending upwards towards the head, is called the __________________ artery.

---

right vertebral

1179. The blood from the right spermatic or ovarian vein empties into the ______________.

---

inferior vena cava

1180. The ______________ arteries extend from the abdominal aorta to the muscles of the skin and abdominal wall, to the lumbar vertebrae, and to the spinal cord.

---

lumbar

1181. From the muscles of the skin and abdominal wall, from the lumbar vertebrae and from the spinal cord the blood flows into the ______________ veins

---

lumbar

1182. into the ______________.

---

inferior vena cava

1183. The ______________ artery extends from the abdominal aorta to the sacrum and coccyx.

---

middle sacral

1184. From the sacrum and coccyx the blood flows into the ______________ vein

---

middle sacral

1185. into the ______________.

---

inferior vena cava

1186. At its end the abdominal aorta divides into the right and left ______________ arteries.

---

common iliac

1187. The right and left common iliac arteries subdivide into the ______________ arteries and the ______________ arteries.

---

external iliac, hypogastric

1188. The ______________ arteries extend from the common iliac arteries to the pelvis.

---

hypogastric

1189. From the pelvis the blood flows into the ______________ veins

---

hypogastric

1190. to the ______________ veins

---

common iliac

1191. to the ______________.

---

inferior vena cava

---

235. The corresponding artery branching from the left subclavian artery is called the ________________ artery.

left vertebral

236. The upward branches of the right and left subclavian arteries are called the __________ and ________ ________________ arteries.

right, left vertebral

237. Label the following diagram:

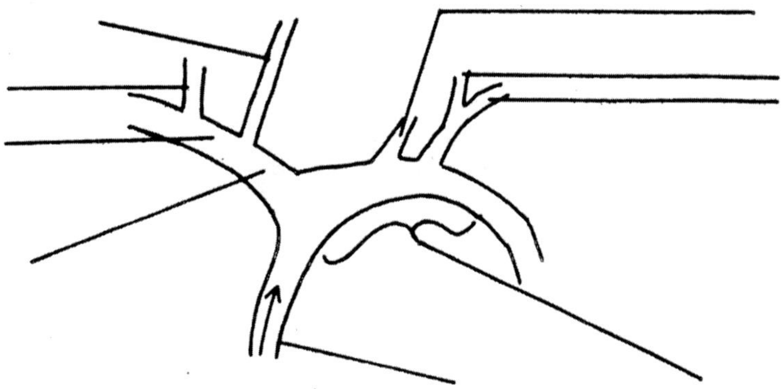

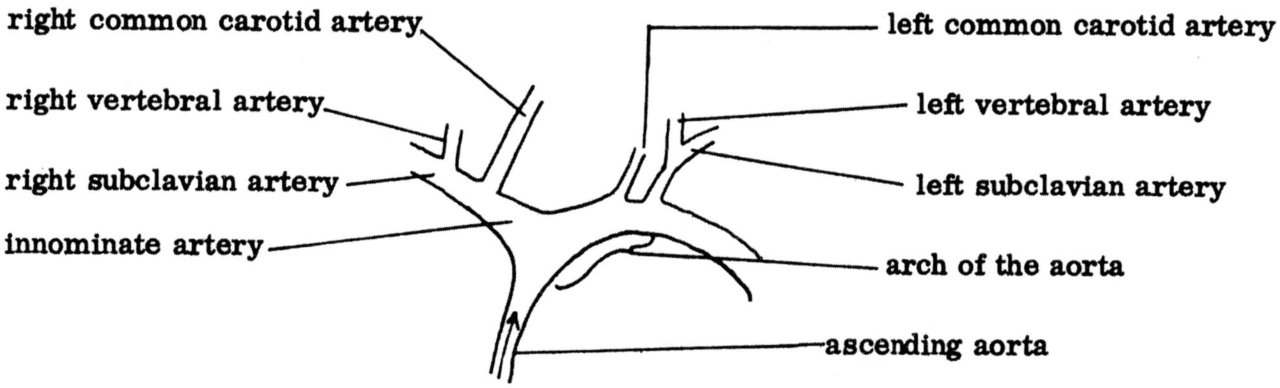

1164. and the ____________ vein carries the blood from there

inferior mesenteric

1165. into the ____________ vein

splenic

1166. and then into the ______________ vein.

portal

1167. The ______________ arteries extend from the abdominal aorta to the kidneys.

right and left renal

1168. From the kidneys the blood flows into the _______________ veins

right and left renal

1169. and then into the _________________.

inferior vena cava

1170. The ______________ artery extends from the abdominal aorta to the suprarenal glands.

right and left suprarenal

1171. From the suprarenal glands the blood flows into the ________________ veins

right and left suprarenal

1172. From the left suprarenal vein the blood flows into the ______________ vein

left renal

1173. and then into the _________________.

inferior vena cava

1174. From the right suprarenal vein the blood flows into the ________________.

inferior vena cava

1175. The _____________ arteries extend from the abdominal aorta to the testes or ovaries.

right and left spermatic or ovarian

1176. From the testes or ovaries the blood flows into the ________________ veins.

right and left spermatic or ovarian

1177. The blood from the left spermatic or ovarian vein flows into the _____________ vein

left renal

1178. and then into the _______________.

inferior vena cava

IV. THE UPPER EXTREMITIES

238. The subclavian arteries supply blood to the upper extremities.

239. The right subclavian artery supplies blood to the ______ arm.

right

240. The left subclavian artery furnishes blood to the ______ arm.

left

241. At the outer border of the first rib the subclavian artery becomes the axillary artery.

242. The right and left subclavian arteries continue until the outer border of the ______ rib.

first

243. After passing the outer border of the first rib the subclavian artery is called the ______ artery.

axillary

244. The ______ artery is a continuation of the subclavian artery, beginning after the first rib.

axillary

245. Label the following diagram:

246. After the subclavian artery passes the first rib it is called the axillary artery. After the axillary artery passes the axilla (armpit) it is called the brachial artery.

247. The continuation of the axillary artery is called the ______ artery.

brachial

248. The axillary artery continues from the outer edge of the first rib to the axilla, after which it is called the ______ artery.

brachial

1150. From there the blood flows into the ________________ vein

right gastric

1151. into the ____________ vein.

portal

1152. The ______________ artery carries the blood from the hepatic artery to the stomach and duodenum.

gastroduodenal

1153. From the stomach and duodenum the blood flows into the ___________ vein,

gastroduodenal

1154. and then into the ______________ vein.

portal

1155. The __________ artery carries the blood from the hepatic artery to the gallbladder.

cystic

1156. The ____________ vein carries the blood from the gallbladder

cystic

1157. to the ____________ vein.

portal

1158. From the portal vein the blood flows to the _______________,

liver

1159. and then into the _____________ veins to the _____________.

hepatic, inferior vena cava

1160. The ______________ artery extends from the abdominal aorta to the small intestine and to half of the large intestine.

superior mesenteric

1161. From the small intestine and half of the large intestine the blood flows into the _________________ vein

superior mesenteric

1162. to the _______________ vein.

portal

1163. The ______________ artery extends from the abdominal aorta to the other half of the large intestine,

inferior mesenteric

249. Reviewing. The first branch of the arch of the aorta is called the __________ artery.

innominate

250. The innominate artery subdivides into two branches, the ________________________ artery and the _____________________ artery.

right common carotid, right subclavian

251. The subclavian artery continues until the outer edge of the first rib, after which it is called the _______________ artery.

axillary

252. The axillary artery continues from the outer edge of the first rib to the axilla, after which it is called the ________________ artery.

brachial

253. Label the following diagram:

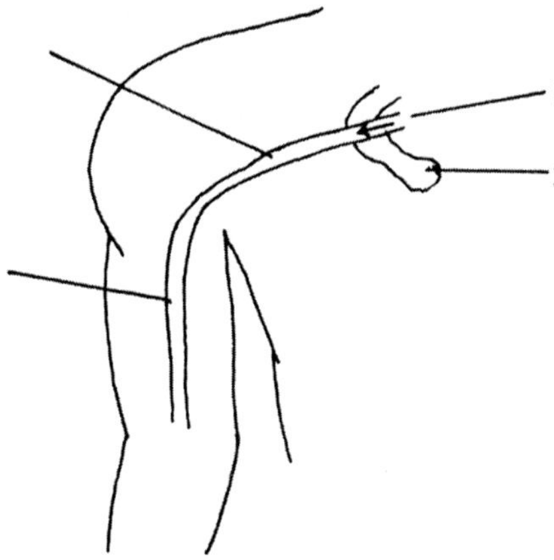

subclavian artery

first rib

axillary artery

brachial artery

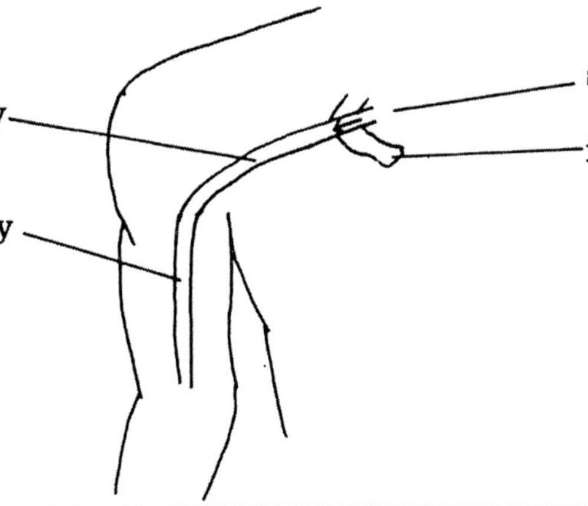

subclavian artery

first rib

254. Just below the elbow the brachial artery divides into two branches, the radial artery and the ulnar artery.

255. The brachial artery divides into two branches, the radial artery and the ulnar artery. This division takes place just below the ________________.

elbow

1136. Time for one final review.

The __________ arteries extend from the abdominal aorta to the under surface of the diaphragm.

---

inferior phrenic

1137. From the under surface of the diaphragm the blood flows into the ____________ veins,

---

inferior phrenic

1138. and then into the ____________.

---

inferior vena cava

1139. The next artery extending from the abdominal aorta is the ____________ artery.

---

celiac

1140. The three branches of the celiac artery are the ___________ artery, the ___________ artery, and the ___________ artery.

---

left gastric, splenic, hepatic

1141. The ____________ artery extends from the celiac artery to the stomach and esophagus.

---

left gastric

1142. From the stomach and esophagus the blood flows into the ____________ vein

---

left gastric

1143. and then into the ______________ vein.

---

portal

1144. The ____________ artery extends from the celiac artery to the stomach, pancreas, and the spleen.

---

splenic

1145. Then the blood flows from the pancreas, stomach and spleen into the ___________ vein

---

splenic

1146. into the ____________ vein.

---

portal

1147. The ____________ artery extends from the celiac artery to the liver.

---

hepatic

1148. The hepatic artery also subdivides into three branches, the ___________ artery, the ___________ artery, and the ___________ artery.

---

right gastric, gastroduodenal, cystic

1149. The ____________ artery extends from the hepatic artery to the stomach.

---

right gastric

256. The brachial artery divides into two branches, the r__________ artery and the ulnar artery.

---

radial

257. The two branches of the brachial artery are the ulnar artery and the ____________ artery.

---

radial

258. The brachial artery divides into two branches. The larger branch is called the ulnar artery. The smaller branch is called the _____________ artery.

---

radial

259. The brachial artery subdivides into two branches. The smaller branch is the radial artery. The larger branch is the _______________ artery.

---

ulnar

260. The two branches of the brachial artery are the ______________ artery and the ____________ artery.

---

radial, ulnar

261. Reviewing, the subclavian artery extends to the outer edge of the first rib, after which it is called the ________________ artery.

---

axillary

262. The axillary artery continues to the axilla, after which it is called the _____________ artery.

---

brachial

263. The brachial artery extends to a point just below the elbow where it subdivides into two branches, the __________ artery and the ____________ artery.

---

radial, ulnar

264. Label the following diagram:

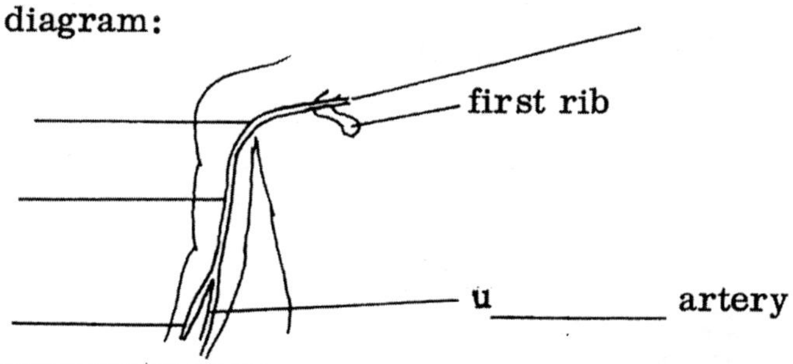

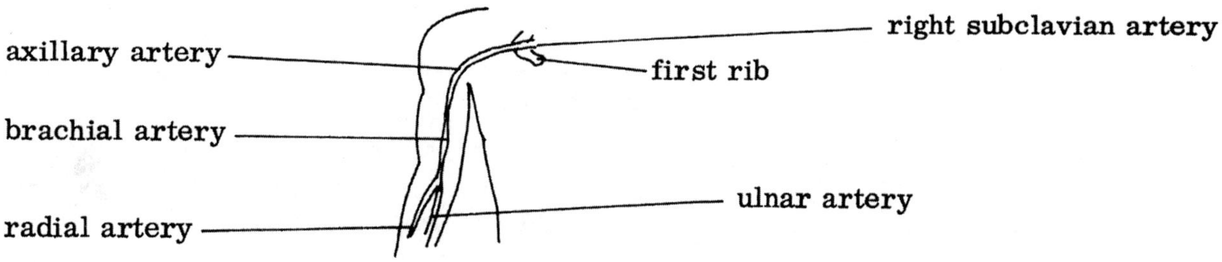

1121. The ____________ veins receive the blood flowing from the pelvis.

---

hypogastric

1122. Blood flows to the pelvis through the _______________ arteries.

---

hypogastric

1123. Blood flows from the pelvis through the _________________ veins.

---

hypogastric

1124. The hypogastric veins empty into the <u>common iliac veins</u>.

1125. The veins which receive the blood flowing from the hypogastric veins are the common __________________ veins.

---

iliac

1126. The _____________ iliac veins receive the blood flowing from the hypogastric veins.

---

common

1127. Blood flows from the hypogastric veins into the _________________ veins.

---

common iliac

1128. Blood flows to the pelvis through the ________________ arteries

---

hypogastric

1129. and from the pelvis through the ________________ veins

---

hypogastric

1130. into the _____________ veins.

---

common iliac

1131. The common iliac veins empty into the <u>inferior vena cava</u> which carries the blood back to the heart.

1132. The blood flowing from the common iliac veins passes into the ______________ which carries it back to the heart.

---

inferior vena cava

1133. The blood flows from the pelvis through the _______________ veins

---

hypogastric

1134. into the ________________ veins

---

common iliac

1135. and then empties into the _______________ which carries it back to the heart.

---

inferior vena cava

---

265. Both the radial and the ulnar arteries extend into the palm of the hand where they subdivide into the superficial and deep volar arches.

---

266. In the palm of the hand the radial and ulnar arteries subdivide into superficial and ________ volar arches.

deep

---

267. The superficial and deep ____________ arches are found in the palm of the hand. They are formed by the subdividing of the ulnar and radial arteries.

volar

---

268. When the radial and ulnar arteries subdivide in the palm of the hand, they form the superficial and ____________________________.

deep volar arches

---

269. The palm of the hand contains the ________________________________ formed by the subdivision of the radial and ulnar arteries.

superficial and deep volar arches

---

270. The superficial and deep volar arches extend into the digital arteries.

---

271. The blood flows from the superficial and deep volar arches into the fingers through the ____________________ arteries.

digital

---

272. The fingers receive their blood supply from the ________________ arteries.

digital

---

273. The superficial and deep volar arches extend into the fingers and are then called the _____________ arteries.

digital

---

274. Label the following diagram:

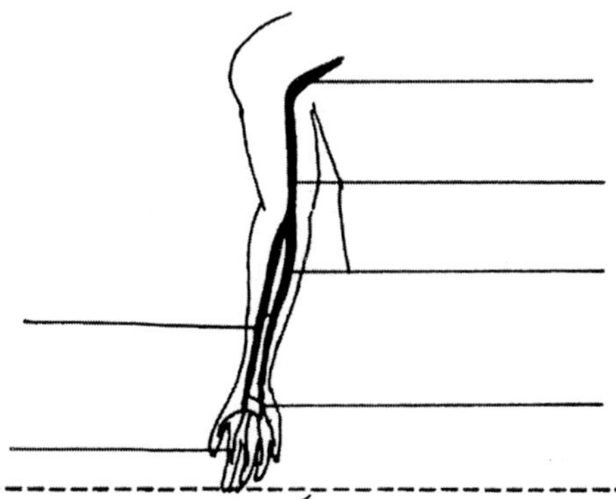

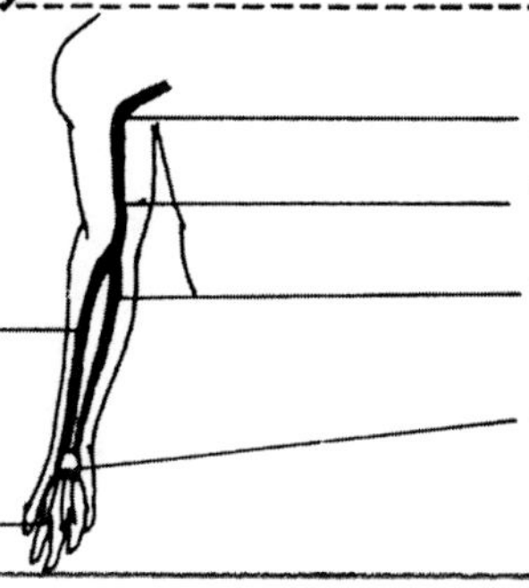

1107. The two terminal branches of the abdominal aorta are the ________________ arteries.

---

right and left common iliac

1108. The right and left common iliac arteries divide into the <u>external iliac arteries</u> and the <u>hypogastric arteries</u> (also known as the internal iliac arteries).

1109. The right common iliac artery divides into the right external iliac artery and the right hypogastric artery. The left common iliac artery divides into the left external ________________ artery and the left hypogastric artery.

---

iliac

1110. The two branches of the common iliac arteries are the ____________ iliac arteries and the hypogastric arteries.

---

external

1111. The common iliac artery subdivides into the external iliac artery and the hypo___________ artery.

---

gastric

1112. One branch of the common iliac artery is the external iliac artery. The other branch is the ___________ artery.

---

hypogastric

1113. The blood flows from the common iliac arteries into the _____________ arteries and the hypogastric arteries.

---

external iliac

1114. The common iliac arteries divide into the _________ arteries and the _________ arteries.

---

external iliac, hypogastric

1115. At its end the abdominal aorta divides into the right and left _____________ arteries.

---

common iliac

1116. The common iliac arteries divide into the ___________ arteries and the _________ arteries.

---

external iliac, hypogastric

1117. The <u>hypogastric artery</u> supplies blood to the <u>pelvis</u>.

1118. The branch of the common iliac artery which supplies blood to the pelvis is the _______________ artery.

---

hypogastric

1119. The hypogastric artery supplies blood to the _____________.

---

pelvis

1120. The blood flows to the pelvis through the hypogastric arteries and from the pelvis through the corresponding veins, the _______________ veins.

---

hypogastric

275. From the digital arteries the blood flows into arterioles and then into capillaries in order to nourish the cells.

---

276. From the digital arteries the blood flows into ____________ and then into the capillaries.

arterioles

---

277. The blood flows from the arteries into the ____________ and then into the ____________.

arterioles, capillaries

---

278. The blood flows from the digital arteries into the ____________ and then into the ____________ to nourish the cells.

arterioles, capillaries

---

279. Reviewing: (If you don't need a review, proceed to frame #290)
The blood supply for the right arm flows through the following path:
Blood flows from the heart into the ____________ aorta

ascending

---

280. and then into the ____________ of the aorta.

arch

---

281. From the arch of the aorta the blood flows through the ____________ artery

innominate

---

282. into the ____________ artery.

right subclavian

---

283. From the right subclavian artery the blood flows into the ____________ artery

axillary

---

284. and then into the ____________ artery

brachial

---

285. which subdivides below the elbow into the ____________ and ____________ arteries.

radial, ulnar

---

286. The radial and ulnar arteries subdivide in the palm of the hand to form the ________________________.

superficial and deep volar arches.

---

287. The blood flows from the superficial and deep volar arches into the ____________ arteries

digital

---

288. then into the ____________

arterioles

---

289. and finally into the ____________ to nourish the cells.

capillaries

1102. Label the indicated parts of the abdominal aorta:

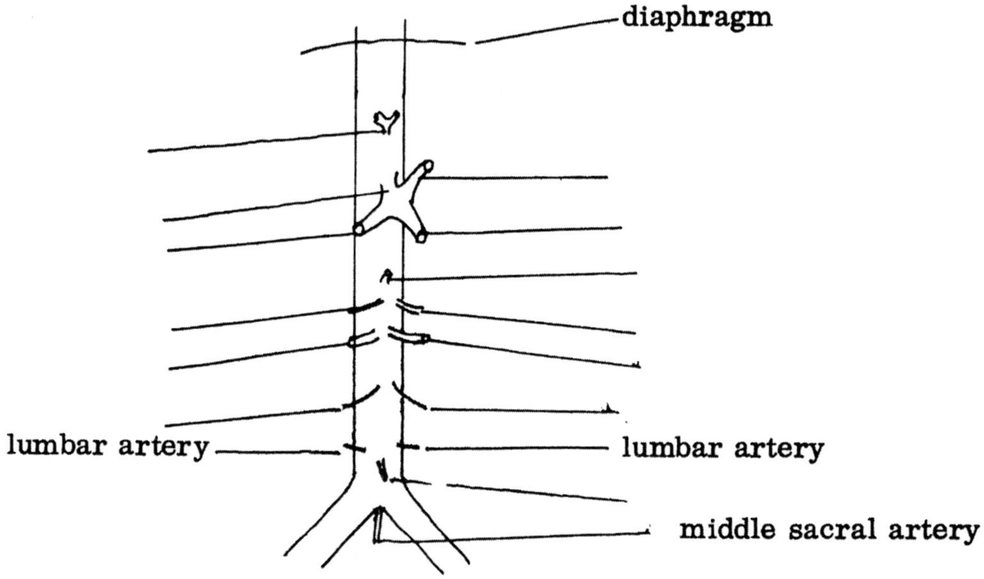

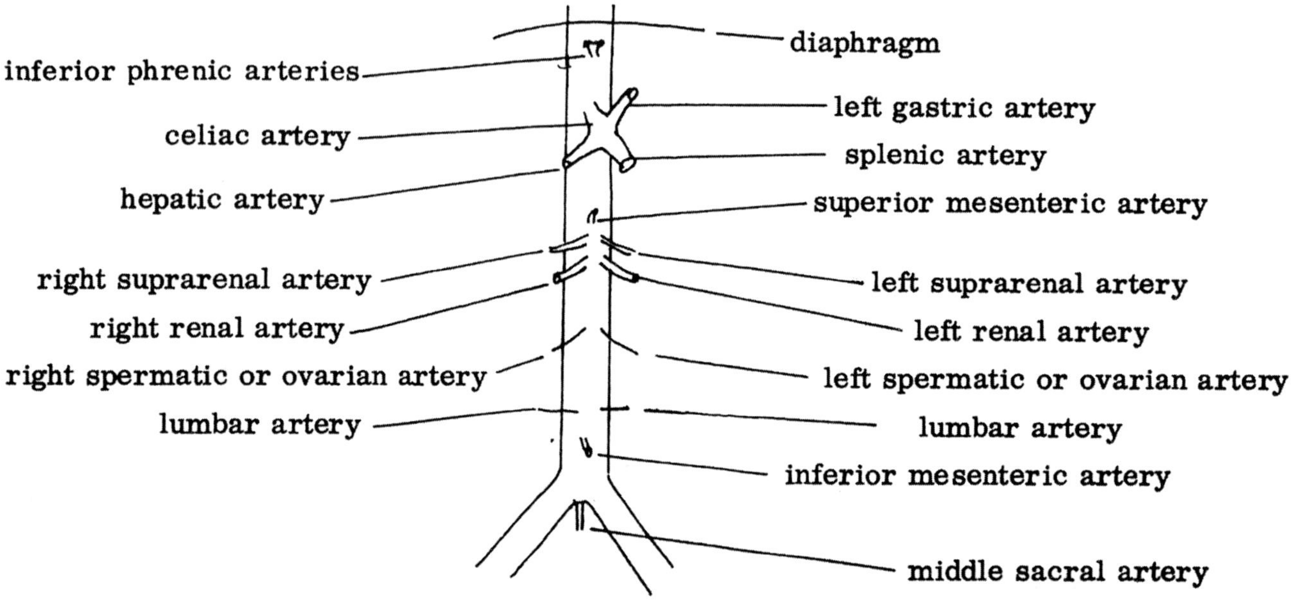

1103. At its lower end the abdominal aorta divides into two branches, the right and left common iliac arteries.

1104. The arteries at the end of the abdominal aorta are the right and left ____________ iliac arteries.

---

common

1105. At its end the abdominal aorta divides into two branches, the right and left common ____________ arteries.

---

iliac

1106. The terminal branches of the abdominal aorta are the right and left ____________ arteries.

---

common iliac

290. Trace the path of the blood from the heart to the capillaries of the left hand.

---

heart to ascending aorta to arch of the aorta to left subclavian artery to axillary artery to brachial artery to radial and ulnar arteries to superficial and deep volar arches to digital arteries to arterioles to capillaries.

---

291. From the capillaries the blood flows into the venules.

---

292. After leaving the capillaries the blood then enters the ____________.

---

venules

---

293. The circulation of the blood is from the arteries to the arterioles to the capillaries to the ____________.

---

venules

---

294. The venules then carry the blood to the superficial and deep volar venous arches.

---

295. From the venules the blood flows into the superficial and deep ____________ ____________ arches.

---

volar venous

---

296. The superficial and ________ ________ ________ arches collect the blood flowing from the venules.

---

deep volar venous

---

297. The blood flows from the capillaries to the venules to the ________________________.

---

superficial and deep volar venous arches

---

298. Review time again. (If you feel that you do not need a review, proceed to frame 312). Blood flows from the heart to the ____________ aorta.

---

ascending

---

299. then to the ____________ of the aorta.

---

arch

---

300. The blood supply for the right arm then flows through the ____________ artery

---

innominate

---

301. to the ____________________ artery.

---

right subclavian

---

302. The blood supply for the left arm flows from the arch of the aorta into the ____________ artery.

---

left subclavian

---

303. From the subclavian artery the blood flows into the ____________ artery

---

axillary

1090. The artery extending from the abdominal aorta to the sacrum and coccyx is the ___________ artery.

middle sacral

1091. The blood flows to the sacrum and coccyx through the ______________ artery.

middle sacral

1092. The blood flows from the sacrum and coccyx through the corresponding vein, the ____________ vein.

middle sacral

1093. The vein which receives the blood from the sacrum and the coccyx is the ___________ vein.

middle sacral

1094. From the middle sacral vein the blood flows into the inferior vena cava and then back to the heart.

1095. The ______________ receives the blood flowing from the sacrum and coccyx.

inferior vena cava

1096. The blood flows to the muscles of the skin and abdominal wall, to the lumbar vertebrae and to the spinal cord through the ______________ arteries

lumbar

1097. and from there through the _____________ veins

lumbar

1098. into the __________.

inferior vena cava

1099. The blood flows to the sacrum and the coccyx through the ______________ artery

middle sacral

1100. and from there into the _____________ vein

middle sacral

1101. and then into the ________________.

inferior vena cava

304. to the ________________ artery.

---

brachial

305. At the elbow the brachial artery subdivides into the _____________ and _______________ arteries

---

radial, ulnar

306. which join in the palm of the hand to form the ____.

---

superficial and deep volar arches

307. From the superficial and deep volar arches the blood flows into the _____________ arteries

---

digital

308. to the ___________________

---

arterioles

309. and then to the __________________.

---

capillaries

310. From the capillaries the blood flows into the ______________

---

venules

311. and then into the ____.

---

superficial and deep volar venous arches

312. When the blood reaches the superficial and deep volar venous arches, it has two different paths to follow--one through the superficial veins (just below the skin) and the other through the deep veins.

---

313. The blood from the superficial volar venous arches flows into a network of superficial veins, the dorsal venous network of the hand.

---

314. The superficial veins which receive the blood flowing from the superficial volar venous arches are called the dorsal _____________ network of the hand.

---

venous

315. The network through which the blood flows after passing through the superficial volar venous arches is called the ________________ venous network.

---

dorsal

316. The superficial network in the hand is called the _________________________network.

---

dorsal venous

317. The blood flows from the superficial volar venous arches into the _____________________ of the hand.

---

dorsal venous network

318. From the dorsal venous network two large veins extend superficially upward, returning the blood towards the heart. They are the cephalic vein and the basilic vein.

1077. Trace the path of the blood from the heart to the testes or ovaries and return.

From heart to ascending aorta to arch of aorta to thoracic aorta to abdominal aorta to right and left spermatic or ovarian arteries to testes or ovaries to right and left spermatic or ovarian veins. From left spermatic or ovarian vein to left renal vein to inferior vena cava and back to heart and from right spermatic or ovarian vein to inferior vena cava and back to heart.

1078. Trace the path of the blood from the heart to the suprarenal glands and return.

From heart to ascending aorta to arch of aorta to thoracic aorta to abdominal aorta to right and left suprarenal arteries to suprarenal glands to right and left suprarenal veins. From left suprarenal vein to left renal vein to inferior vena cava to heart and from right suprarenal vein to inferior vena cava to heart.

1079. Just below the spermatic or ovarian arteries on the abdominal aorta are the lumbar arteries, the arteries extending to the muscles of the skin and abdominal wall, to the lumbar vertebrae and to the spinal cord.

1080. The arteries extending from the abdominal aorta to the muscles of the skin and abdominal wall, to the lumbar vertebrae, and to the spinal cord are the ________ arteries.

---

lumbar

1081. The ________ arteries extend from the abdominal aorta to the muscles of the skin and abdominal wall and the corresponding veins, the ________ veins extend from there.

---

lumbar, lumbar

1082. The ________ veins receive the blood flowing from the lumbar vertebrae, from the spinal cord and from the muscles of the skin and abdominal wall.

---

lumbar

1083. The lumbar veins empty into the inferior vena cava which carries the blood back to the heart.

1084. The blood flows to the muscles of the skin and abdominal wall, to the lumbar vertebrae and to the spinal cord through the ________ arteries

---

lumbar

1085. and from there into the ________ veins

---

lumbar

1086. which empty into the ________.

---

inferior vena cava

1087. Just above the end of the abdominal aorta is the middle sacral artery, the artery extending to the sacrum and coccyx (kok' siks).

1088. The artery extending to the sacrum and coccyx is the middle ________ artery.

---

sacral

1089. The ________ sacral artery leads to the sacrum and the coccyx.

---

middle

319. The cephalic vein extends superficially from the dorsal venous network along the thumb side of the arm.

---

320. The vein which extends superficially from the dorsal venous network along the thumb side of the arm is the ____________ vein.

cephalic

321. The ____________ vein extends superficially along the thumb side of the arm.

cephalic

322. The dorsal venous network extends superficially upward through two large veins, the basilic vein and the ______________ vein.

cephalic

323. The basilic vein extends superficially upward along the little finger side of the arm.

---

324. The vein which extends superficially along the little finger side of the arm is the ______________ vein.

basilic

325. The vein which extends superficially along the thumb side of the arm is the __________ vein.

cephalic

326. The ____________ vein extends superficially upward along the little finger side of the arm.

basilic

327. The dorsal venous network extends superficially upwards through two large veins, the ____________ and the ______________.

cephalic, basilic

328. From the capillaries the blood flows into the ______________.

venules

329. From the venules the blood flows into the ____.

superficial and deep volar venous arches

330. The blood flows from the superficial volar venous arches into the ____________________.

dorsal venous network of the hand

331. The dorsal venous network branches into the __________ and __________ veins.

cephalic, basilic

332. The cephalic vein extends superficially upward to empty into the axillary vein.

---

333. The vein which the cephalic vein empties into is the ________________ vein.

axillary

1071. The ____________ arteries extend from the abdominal aorta to the testes or ovaries.

right and left spermatic or ovarian

1072. From the testes or the ovaries the blood flows into the __________________ veins.

right and left spermatic or ovarian

1073. From the left spermatic or ovarian vein the blood flows into the ______________ vein,

left renal

1074. and then into the ________________.

inferior vena cava

1075. The right spermatic or ovarian vein empties into the ____________________.

inferior vena cava

1076. Label the following parts of the abcominal aorta:

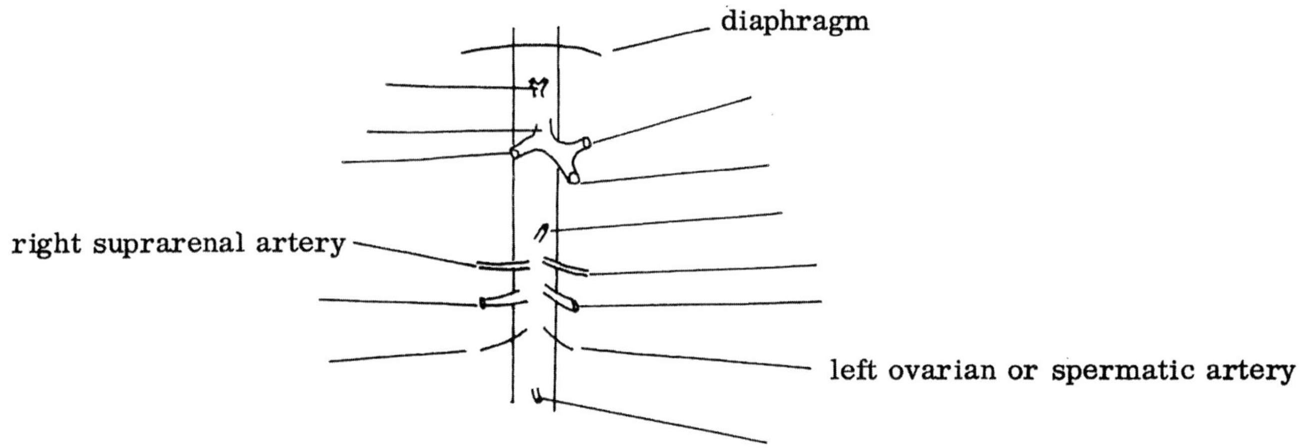

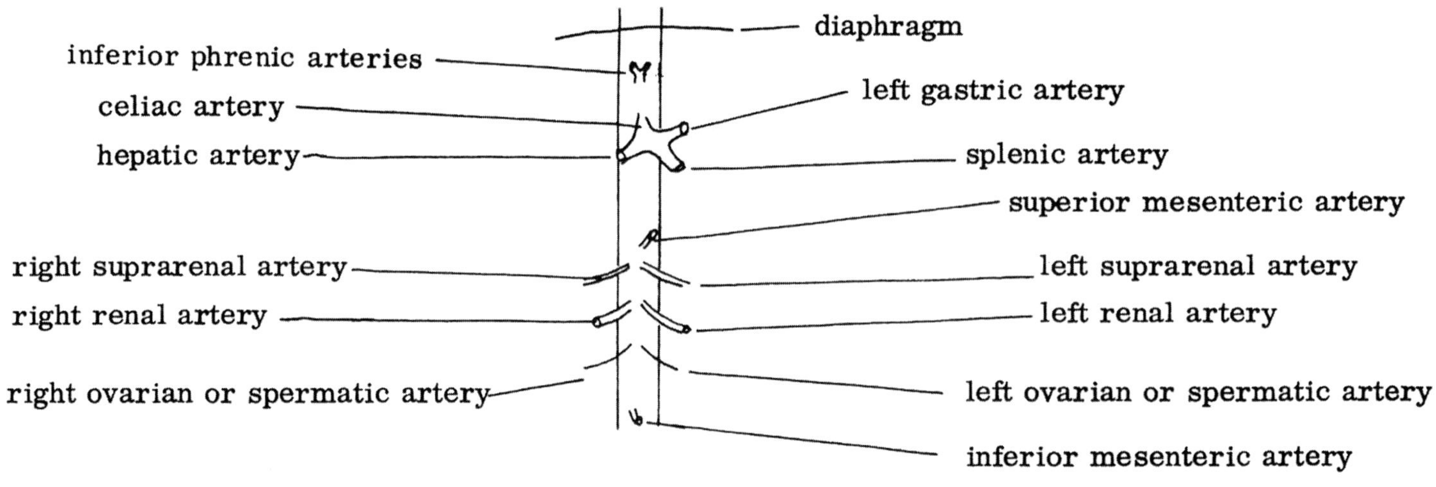

334. The ______________ vein receives the blood flowing from the cephalic vein.

axillary

335. The cephalic vein extends superficially upward and empties into the ____________________.

axillary vein

336. The basilic vein extends superficially upward and then joins the brachial vein to form the axillary vein.

337. The basilic vein extends superficially upward and then joins the ______________ vein.

brachial

338. The _______________ vein receives the blood flowing from the basilic vein.

brachial

339. The joined brachial and basilic veins form the ____________________ vein.

axillary

340. The ________________ vein is formed by the combination of the brachial and the basilic veins.

axillary

341. The basilic vein extends superficially upward to join the _______________ vein and then form the ____________________ vein.

brachial, axillary

342. Time for another review. (If you don't need one, proceed to frame 360).
Blood flows from the heart into the _________________ aorta

ascending

343. into the _______________ of the aorta.

arch

344. The blood for the right arm then flows into the _________________ artery

innominate

345. and then into the ________________ artery.

right subclavian

346. The blood supply for the left arm flows through the ___________________ artery.

left subclavian

347. From the subclavian artery the blood flows into the ______________ artery

axillary

348. into the ______________ artery.

brachial

1057. which leads to the ________________.

inferior vena cava

1058. The blood flowing from the right spermatic vein or the right ovarian vein flows directly into the inferior vena cava.

1059. The ________________ receives the blood flowing from the right spermatic or ovarian vein.

inferior vena cava

1060. The ____________ receives the blood flowing from the left spermatic or ovarian vein.

left renal vein

1061. Time for another review (or else go on to frame 1076)
The ______________ arteries extend from the abdominal aorta to the under surface of the diaphragm.

inferior phrenic

1062. The ____________ artery extends from the abdominal aorta to the stomach, spleen, gall-bladder, pancreas, and liver.

celiac

1063. The _____________ artery extends from the abdominal aorta to the small intestine and to half of the large intestine.

superior mesenteric

1064. The ________________ artery extends from the abdominal aorta to the other half of the large intestine.

inferior mesenteric

1065. The ________________ arteries extend from the abdominal aorta to the kidneys.

right and left renal

1066. The ________________ arteries extend from the abdominal aorta to the suprarenal glands.

right and left suprarenal

1067. From the right and left suprarenal glands the blood flows into the ________________ veins.

right and left suprarenal

1068. From the left suprarenal vein the blood flows into the _____________ vein

left renal

1069. and then into the ________________.

inferior vena cava

1070. The blood from the right suprarenal vein flows into the __________________.

inferior vena cava

349. Just below the elbow the brachial artery subdivides into the ______________ and ______________ arteries.

radial, ulnar

350. The radial and ulnar arteries join in the palm of the hand to form the ________________

superficial and deep volar arches

351. which lead to the __________________ arteries.

digital

352. From the digital arteries the blood flows into the ____________________

arterioles

353. and then into the _________________ to nourish the cells.

capillaries

354. From the capillaries the blood enters the _____________

venules

355. and then the ________________________________

superficial and deep volar venous arches

356. From the superficial volar venous arches the blood flows into the ______________________ network of the hand.

dorsal venous

357. The dorsal venous network subdivides into two large veins, the _________________ and ____________________.

cephalic, basilic

358. The cephalic vein extends superficially upward and empties into the ______________ vein.

axillary

359. The basilic vein extends superficially upward and joins the _______________ vein to form the ______________ vein.

brachial, axillary

360. The blood from the venules enters the superficial and deep volar venous arches. The superficial volar venous arches branch into the cephalic vein and the basilic vein. The deep volar venous arches branch into the radial and ulnar veins.

361. The radial and ______________ veins branch from the deep volar venous arches.

ulnar

362. The deep volar venous arches subdivide into the ________________ and ulnar veins.

radial

1042. The right and left ovarian arteries extend to the ______________.

ovaries

1043. The arteries extending from the abdominal aorta to the testes or ovaries are the right and left ________________ arteries.

spermatic or ovarian

1044. From the testes or ovaries the blood flows through the corresponding veins, the right and left _________________________ veins.

spermatic or ovarian

1045. The blood flows to the testes through the __________________ arteries.

right and left spermatic

1046. The blood flows from the testes through the _________________ veins.

right and left spermatic

1047. The blood flows to the ovaries through the _________________ arteries

right and left ovarian

1048. and from the ovaries through the _________________ veins.

right and left ovarian

1049. Both the left spermatic vein and the left ovarian vein empty into the left renal vein.

1050. The vein which receives the blood flowing from the left spermatic vein is the __________ renal vein.

left

1051. The vein which receives the blood flowing from the left ovarian vein is the left __________ vein.

renal

1052. The ______________ vein receives the blood flowing from the left spermatic or the left ovarian vein.

left renal

1053. As before, the blood flows from the left renal vein into the _________________.

inferior vena cava

1054. The blood flows to the right and left testes or ovaries through the ______________ arteries.

right and left spermatic or ovarian

1055. The blood flows from the ovaries or testes through the _____________ veins.

right and left ovarian or spermatic

1056. The blood flows from the left spermatic vein or from the left ovarian vein into the _____________ vein

left renal

363. The two veins which lead from the deep volar venous arches are the ______________ and ______________ veins.

radial, ulnar

364. The radial and ulnar veins combine just below the elbow to form the brachial vein.

365. Just below the elbow the radial and ulnar veins combine to form the _____________ vein.

brachial

366. The _________________ vein receives the blood flowing from the radial and ulnar veins.

brachial

367. The brachial vein extends upwards into the axillary vein.

368. From the brachial vein the blood enters the ________________ vein.

axillary

369. The _______________ vein receives the blood flowing from the brachial vein.

axillary

370. The axillary vein also receives the blood flowing from the cephalic vein as well as from the basilic vein which joined the brachial vein.

371. Reviewing the superficial and deep venous system of the arm, the blood flows from the venules into the _____________________________________.

superficial and deep volar venous arches

372. From the superficial volar venous arches the blood enters the __________________________.

dorsal venous network of the hand

373. The dorsal venous network extends into two branches, the _________________ and the _______________ veins.

cephalic, basilic

374. The cephalic vein extends superficially upward on the thumb side of the arm and empties into the _________________ vein.

axillary

375. The basilic vein extends superficially upward on the little finger side of the arm and joins the ___________________ vein

brachial

376. to form the ____________________ vein.

axillary

377. The deep volar venous arches subdivide into the __________ and __________ veins,

radial, ulnar

1035. Label the following parts of the abdominal aorta:

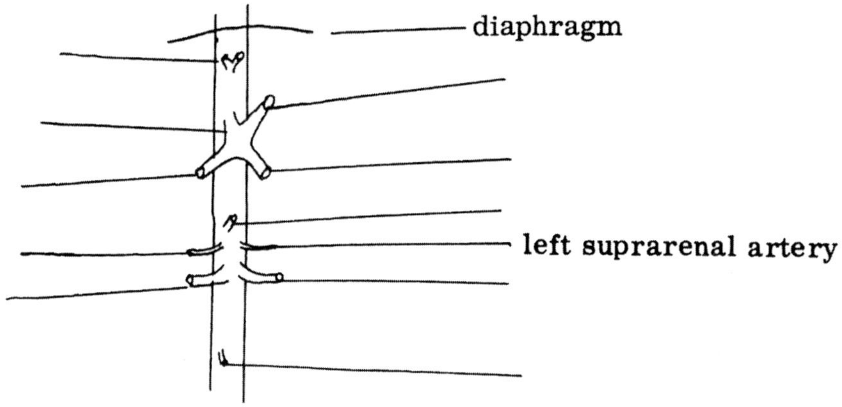

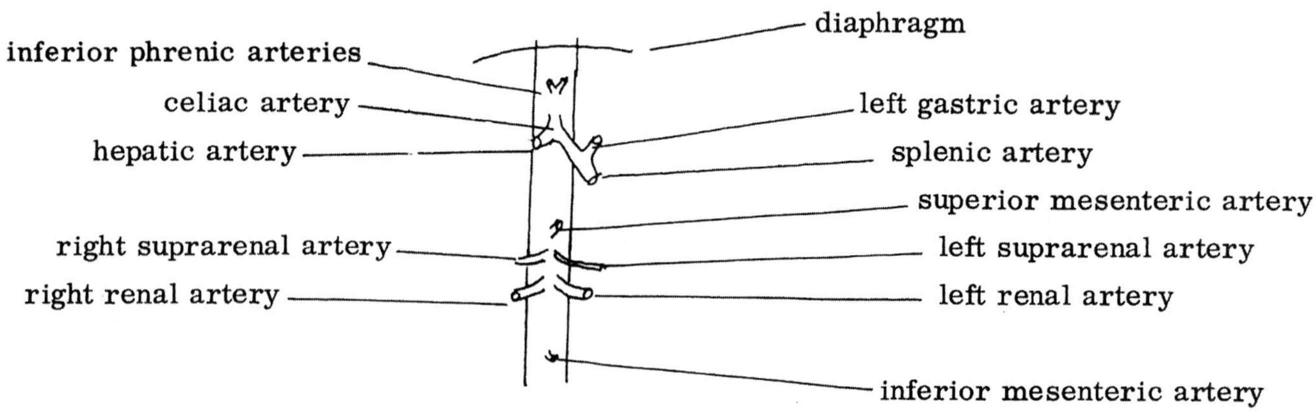

1036. Another pair of arteries arising from the abdominal aorta at a point below the renal arteries are the right and left spermatic or ovarian arteries.

1037. The spermatic arteries extend to the testes and the ovarian arteries to the ____________.

ovaries

1038. The arteries extending to the ovaries are the right and left ______________ arteries.

ovarian

1039. The right and left spermatic arteries extend to the testes and the right and left ovarian arteries extend to the ________________.

ovaries

1040. The right and left ________________ arteries extend to the testes.

spermatic

1041. The right and left spermatic arteries extend to the ________________.

testes

378. which combine just below the elbow to form the ______________ vein

---

brachial

---

379. which in turn empties into the ________________ vein.

---

axillary

---

380. On the right side the axillary vein extends into the right subclavian vein. On the left side the axillary vein extends into the ______________ vein.

---

left subclavian

---

381. The axillary vein extends into the ___________________ vein.

---

subclavian

---

382. The __________________ vein receives the blood flowing from the axillary vein.

---

subclavian

---

383. The right subclavian vein empties into the right innominate vein. The left subclavian vein empties into the left _________________ vein.

---

innominate

---

384. The subclavian veins empty into the __________________ veins.

---

innominate

---

385. The ________________________ vein receives the blood flowing from the left subclavian vein.

---

left innominate

---

386. The _________________ vein receives the blood flowing from the right subclavian vein.

---

right innominate

---

387. The right and left innominate veins empty into the superior vena cava.

---

388. The innominate veins empty into the superior ________ ________.

---

vena cava

---

389. The ______________ vena cava receives the blood flowing from the right and left innominate veins.

---

superior

---

390. The right and left innominate veins empty into the _________________________.

---

superior vena cava

---

391. The superior vena cava extends into the right atrium of the heart.

---

392. From the superior vena cava the blood flows into the right ____________ of the heart.

---

atrium

1024. and then into the ______________ vein.

left renal

1025. From the left renal vein the blood flows into the inferior vena cava.

1026. The ______________ receives the blood flowing from the left renal vein.

inferior vena cava

1027. The blood flowing to the right suprarenal gland passes through the ______________ artery,

right suprarenal

1028. and then into the ______________ vein.

right suprarenal

1029. The right suprarenal vein empties directly into the inferior vena cava.

1030. The ______________ receives the blood flowing from the right suprarenal vein.

inferior vena cava

1031. From the left suprarenal vein the blood flows into the ______________ vein

left renal

1032. and then into the ______________.

inferior vena cava

1033. From the right suprarenal vein the blood flows into the ______________.

inferior vena cava

1034. Complete the following table:

abdominal aorta → ________ artery → right suprarenal gland → ____________ vein → inferior vena cava

abdominal aorta → ________ artery → left suprarenal gland → ________ vein → ________ vein → inferior vena cava

abdominal aorta → right suprarenal artery → right suprarenal gland → right suprarenal vein → inferior vena cava

abdominal aorta → left suprarenal artery → left suprarenal gland → left suprarenal vein → left renal vein → inferior vena cava

393. The __________ atrium of the heart receives the blood flowing from the superior vena cava.

---

right

394. From the superior vena cava the blood flows into the ______________ of the heart.

---

right atrium

395. Time for one final review. (If you don't need one, proceed to frame 420).
The path of the blood from the heart to the right hand return is as follows: The blood leaves the left ventricle of the heart and enters the __________ aorta

---

ascending

396. and then flows into the __________ of the aorta.

---

arch

397. From the arch of the aorta the blood flows into the ____________ artery

---

innominate

398. into the ____________ artery

---

right subclavian

399. and then into the ______________ artery.

---

axillary

400. From the axillary artery the blood flows into the ______________ artery

---

brachial

401. which subdivides into the __________ and __________ arteries just below the elbow.

---

radial, ulnar

402. The radial and ulnar arteries combine to form the ______________________________,

---

superficial and deep volar arches

403. from which the blood flows into the ____________ arteries

---

digital

404. into the ______________

---

arterioles

405. and then into the ________________ .

---

capillaries

406. From the capillaries the blood enters the ______________

---

venules

407. and then the ____________________________ .

---

superficial and deep volar venous arches

1009. and then into the ______________ and back to the heart.

inferior vena cava

1010. Just above the renal arteries are the suprarenal arteries, the arteries leading to the suprarenal (adrenal) glands.

1011. The ______________ arteries lead to the suprarenal glands.

suprarenal

1012. The right and left suprarenal arteries lead to the right and left ______________ glands.

suprarenal

1013. The __________ and __________ suprarenal arteries lead to the right and left suprarenal glands.

right, left

1014. The arteries leading to the suprarenal (adrenal) glands are the ______________ arteries.

suprarenal

1015. From the right and left suprarenal glands the blood flows into the corresponding veins, the right and left ______________ veins.

suprarenal

1016. The blood flows to the suprarenal glands through the ______________ arteries.

suprarenal

1017. The blood flows from the suprarenal glands through the ______________ veins.

suprarenal

1018. The veins receiving the blood flowing from the right and left suprarenal glands are the ______________ veins.

right and left suprarenal

1019. From the left suprarenal vein the blood flows into the left renal vein.

1020. The vein which receives the blood flowing from the left suprarenal vein is the left ______________ vein.

renal

1021. The ______________ renal vein receives the blood flowing from the left suprarenal vein.

left

1022. The left suprarenal gland receives the blood flowing from the ______________ artery.

left suprarenal

1023. From the left suprarenal gland the blood flows into the ______________ vein

left suprarenal

408. From the superficial volar venous arches the blood enters the ________________ network of the hand.

---

dorsal venous

409. The two large branches of the dorsal venous network are the ________ and the ________ veins.

---

cephalic, basilic

410. The cephalic vein extends superficially upward along the thumb side of the arm and empties into the ________________ vein.

---

axillary

411. The basilic vein extends superficially upward along the little finger side of the arm and joins the ________________ vein

---

brachial

412. to empty into the ________________ vein.

---

axillary

413. The deep volar venous arches subdivide into two veins, the ________ and the ________.

---

radial, ulnar

414. The radial and ulnar veins combine just below the elbow to form the ________________ vein,

---

brachial

415. which extends into the ________________ vein.

---

axillary

416. From the axillary vein the blood flows into the ____________________ vein

---

right subclavian

417. and then the ____________________________ vein.

---

right innominate

418. From the right innominate vein the blood flows into the _________________________

---

superior vena cava

419. and then into the ________________________ of the heart.

---

right atrium

420. Trace the path of the blood from the heart to the <u>left</u> hand and return.

---

995. From the gallbladder the blood flows into the ________________ vein

cystic

996. and then into the _________________ vein.

portal

997. The portal vein carries the blood to the ____________________.

liver

998. From the liver the blood flows into the _______________ veins

hepatic

999. which empty into the ________________.

inferior vena cava

1000. The branch of the abdominal aorta which leads to the small intestine and half of the large intestine is the _______________ artery.

superior mesenteric

1001. From the small intestine and part of the large intestine the blood flows through the ____________ vein

superior mesenteric

1002. into the _____________ vein.

portal

1003. The branch of the abdominal aorta which extends to the other half of the large intestine is the _____________ artery.

inferior mesenteric

1004. From there the blood flows into the ____________ vein

inferior mesenteric

1005. which carries the blood into the ______________ vein

splenic

1006. and then into the ______________ vein.

portal

1007. Just below the superior mesenteric artery on the abdominal aorta are the right and left ____________ arteries which carry the blood to the kidneys.

renal

1008. From the kidneys the blood flows into the _______________ veins

renal

Left ventricle of heart, ascending aorta, left subclavian artery from arch of aorta, axillary artery, brachial artery, radial and ulnar arteries, superficial and deep volar arches, digital arteries, arterioles, capillaries, venules, superficial and deep volar arches (from superficial volar venous arches to dorsal venous network to cephalic vein to axillary vein and from basilic vein to brachial vein to axillary vein), (from deep volar venous arches to radial and ulnar veins to brachial vein, to axillary vein), left subclavian vein, left innominate vein, superior vena cava, right atrium of the heart.

---

421. **Label the following diagrams:**

Arterial Circulation | Venous circulation | Deep

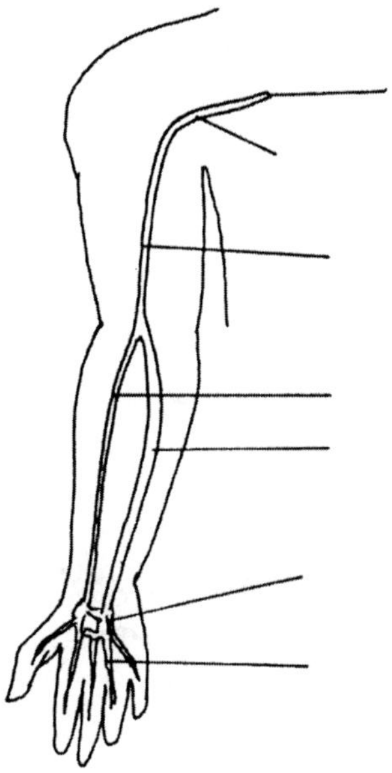

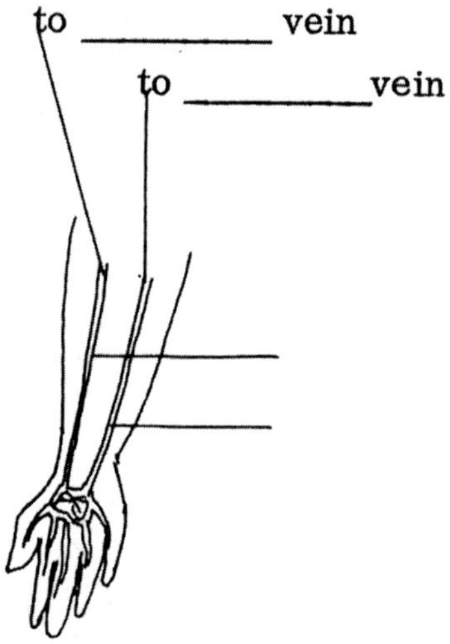

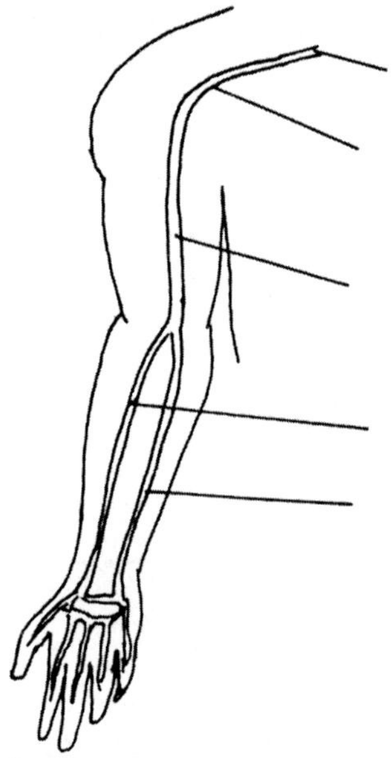

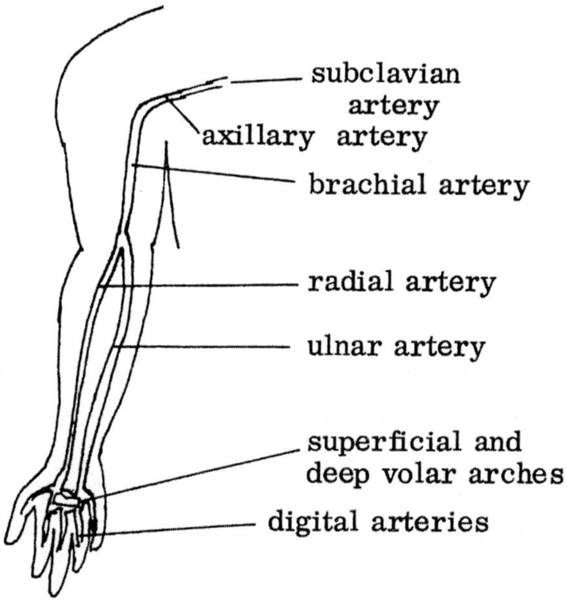

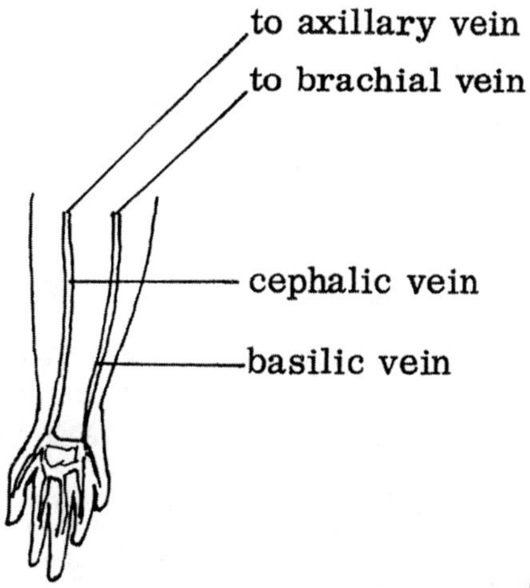

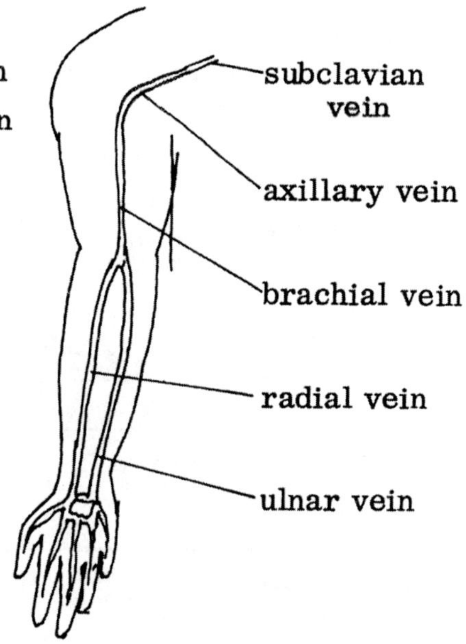

980. The ______________ artery extends to parts of the stomach and esophagus.

left gastric

981. From parts of the stomach and esophagus the blood flows into the ______________ vein

left gastric

982. and then into the ______________ vein.

portal

983. The ______________ artery extends to the stomach, pancreas, and spleen.

splenic

984. From these organs the blood flows into the ______________ vein

splenic

985. and into the ______________ vein.

portal

986. The ______________ artery extends a branch directly to the liver.

hepatic

987. The hepatic artery then subdivides into three branches, the ____________ artery, the ____________ artery, and the ____________ artery.

right gastric, gastroduodenal, cystic

988. The ______________ artery extends from the hepatic artery to the stomach.

right gastric

989. From there the blood flows into the ______________ vein

right gastric

990. and then into the ______________ vein.

portal

991. The ______________ artery carries the blood from the hepatic artery to the stomach and duodenum.

gastroduodenal

992. From the stomach and duodenum the blood flows into the ______________ vein

gastroduodenal

993. which carries the blood to the ______________ vein.

portal

994. The ______________ artery carries the blood from the hepatic artery to the gallbladder.

cystic

V. THE HEAD AND NECK

422. The first artery in the arch of the aorta is the innominate artery. The next artery is the left common carotid artery and the third is the left subclavian artery. Label these on the following diagram:

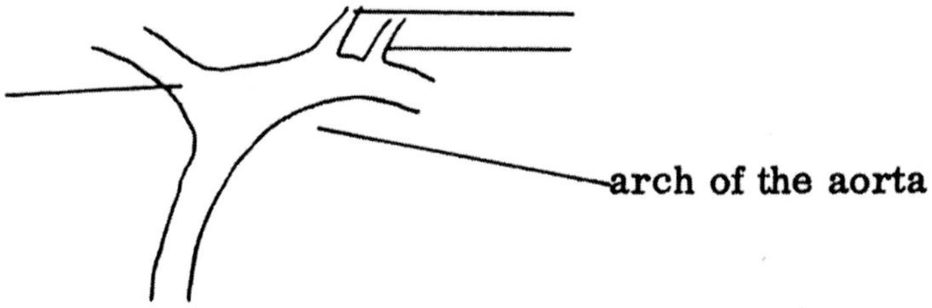

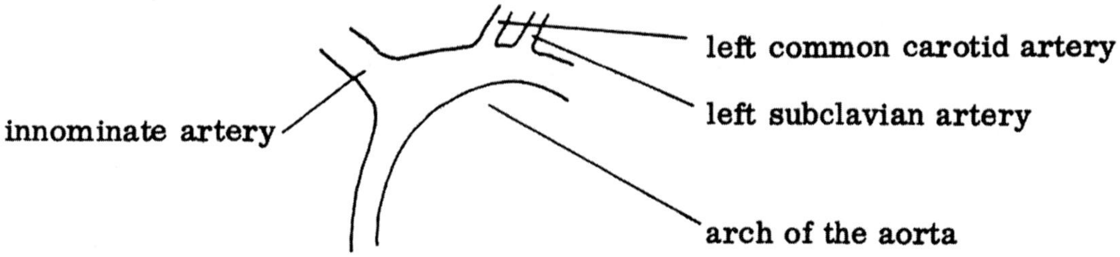

423. The innominate artery subdivides into the right common carotid artery and the right subclavian artery. Each of the subclavian arteries has an upward branch, the vertebral artery.

424. Label the following arteries:

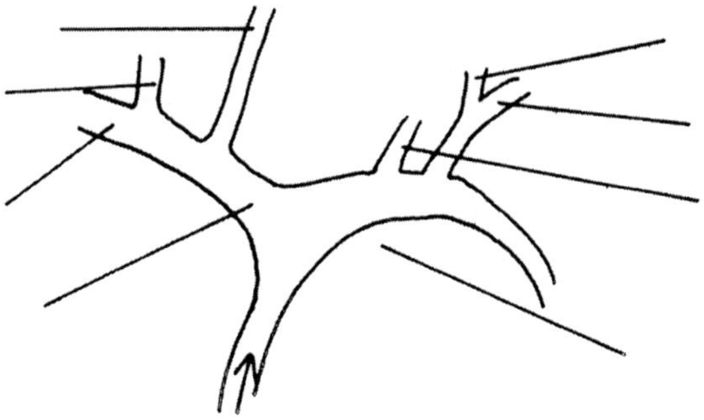

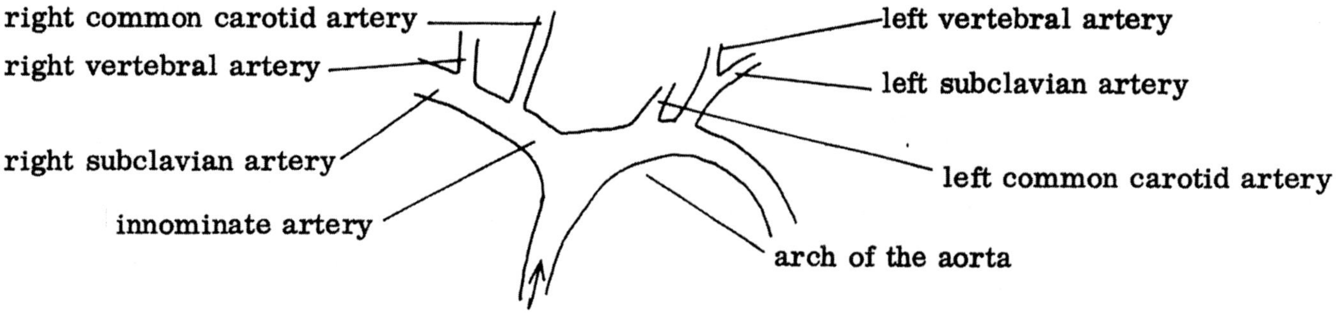

965. The kidneys receive their blood supply from the __________ arteries.

renal

966. The right and left __________ arteries supply blood to the kidneys.

renal

967. The blood flows from the abdominal aorta to the right and left kidneys through the __________ arteries.

right and left renal

968. In the kidneys the blood flows through arterioles, capillaries, venules and then into the corresponding veins, the __________ veins.

renal

969. The veins which receive the blood flowing from the kidneys are the __________ veins.

renal

970. Blood flows to the kidneys through the __________ arteries.

renal

971. Blood flows from the kidneys through the __________ veins.

renal

972. The renal veins empty into the inferior vena cava which carries the blood back to the heart.

973. From the kidneys the blood flows into the __________ veins

renal

974. and then into the __________.

inferior vena cava

975. Time for a big review.
Blood flows from the abdominal aorta into the __________ arteries which carry the blood to the under surface of the diaphragm.

inferior phrenic

976. From the under surface of the diaphragm the blood flows into the __________ veins

inferior phrenic

977. and then into the __________.

inferior vena cava

978. The next artery extending from the abdominal aorta is the __________ artery.

celiac

979. The three branches of the celiac artery are the __________ artery, the __________ artery, and the __________ artery.

left gastric, splenic, hepatic

425. The right and left common carotid arteries and the right and left vertebral arteries extend upwards and supply blood to the head and neck.

---

426. The common carotid arteries (right and left) extend upward along the trachea until they reach the upper border of the thyroid cartilage where they divide into the external carotid artery and the internal carotid artery.

---

427. The common carotid arteries extend upward along the trachea and then divide into the external carotid artery and the ______________ carotid artery.

- -

internal

---

428. When the common carotid arteries reach the upper border of the thyroid cartilage they subdivide into the ____________ and the internal carotid arteries.

- -

external

---

429. The right and left common carotid arteries extend upward along the trachea and subdivide into the ______________________ artery and the __________________ artery.

- -

internal carotid, external carotid

---

430. Label the following diagram:

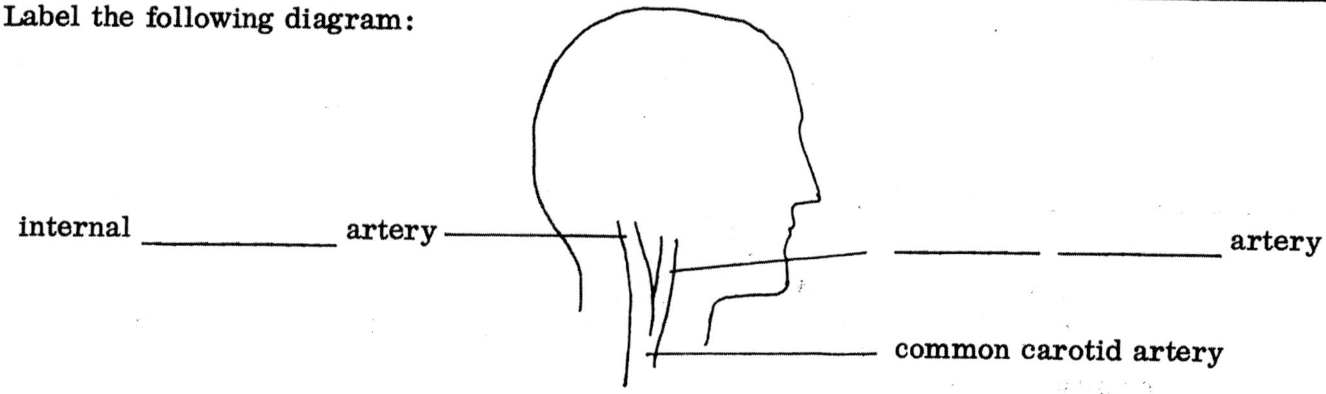

- -

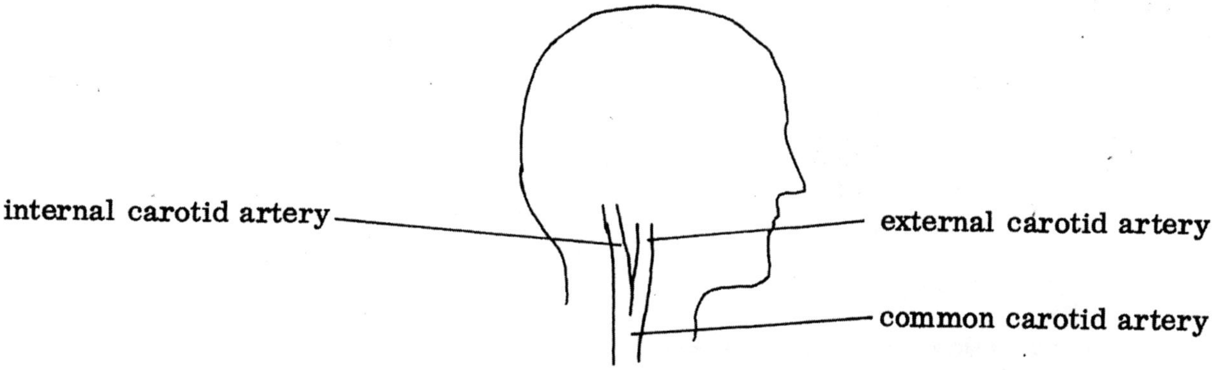

---

431. The external carotid arteries supply blood to the neck, face, mouth, jaws and scalp. The internal carotid arteries supply blood to the brain, eyes, forehead, and nose.

---

432. The external carotid artery has several principal branches, as well as a few minor ones. The first principal branch is the superior thyroid artery which supplies blood to the thyroid gland.

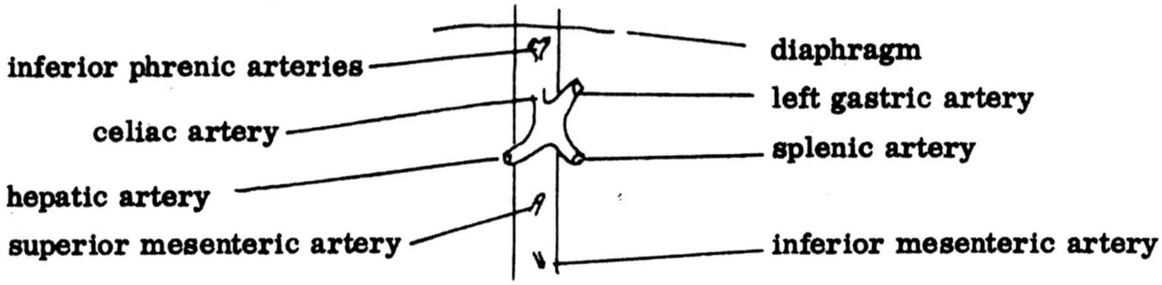

---

961. Along the abdominal aorta, just below the superior mesenteric artery there are two branches, the <u>right and left renal arteries</u> which carry the blood to the <u>kidneys</u>.

---

962. Label the indicated parts of the aorta:

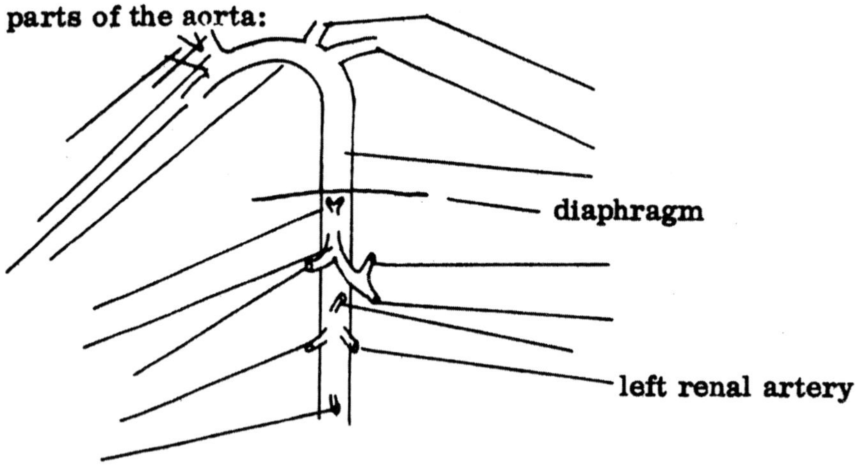

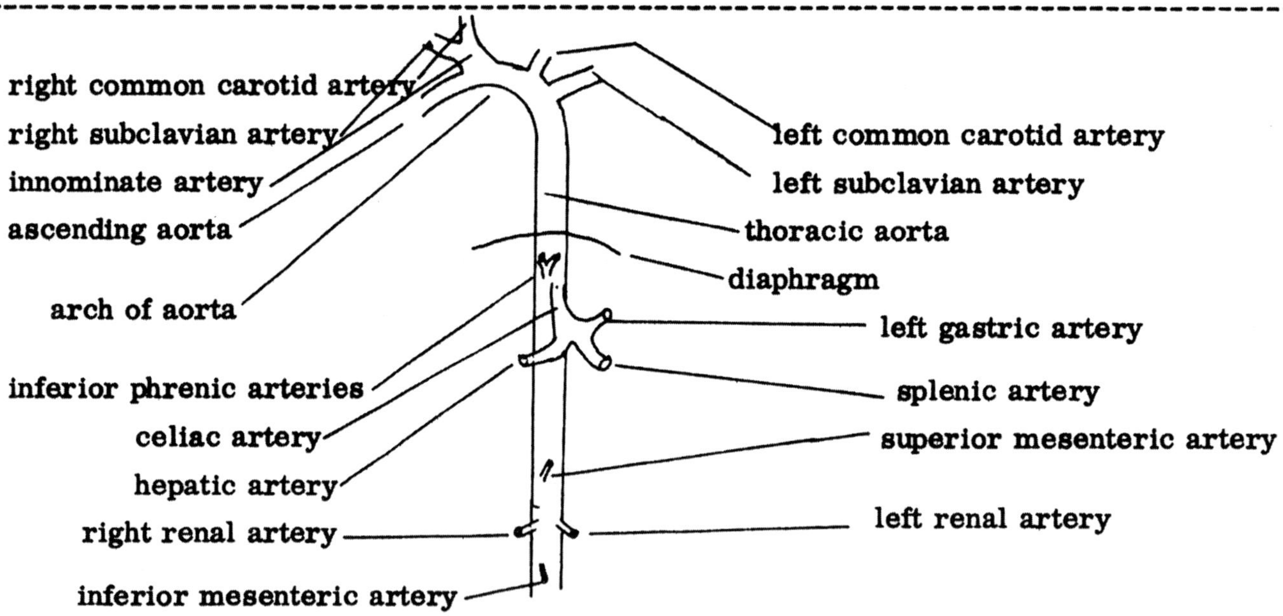

---

963. The right renal artery carries the blood to the right kidney and the ______________ artery carries the blood to the left kidney.

---

left renal

---

964. The right and left renal arteries carry the blood to the right and left ________________.

---

kidneys.

---

433. The first branch of the external carotid artery is the superior ____________ artery, which leads to the thyroid gland.

---

thyroid

---

434. The branch of the external carotid artery which leads to the thyroid gland is the ______________ thyroid artery.

---

superior

---

435. The __________________ artery branches off the external carotid artery and leads to the thyroid gland.

---

superior thyroid

---

436. The first principal artery branching from the external carotid artery is the _______________ artery. It leads to the _______________ gland.

---

superior thyroid, thyroid

---

437. The second principal branch of the external carotid artery is the lingual artery, the artery which leads towards the tongue.

---

438. The artery which branches off the external carotid artery and which leads towards the tongue is the ____________ artery.

---

lingual

---

439. The second principal branch of the external carotid artery is the _____________ artery.

---

lingual

---

440. The _____________ artery branches off the external carotid artery and extends towards the tongue.

---

lingual

---

441. The first principal artery extending from the external carotid artery is the ____________ artery. It leads to the ______________________.

---

superior thyroid, thyroid gland

---

442. The second principal artery branching off the external carotid artery is the ____________ artery, which leads towards the ________________.

---

lingual, tongue

---

951. From the inferior mesenteric vein the blood flows into the ______________ vein

splenic

952. which empties into the ______________ vein.

portal

953. The portal vein carries the blood to the ______________ and from there it flows into the ______________ veins

liver, hepatic

954. carrying the blood to the ____________________ and back to the heart.

inferior vena cava

955. Describe the circulation of the blood from the heart to the small and large intestine and return.

Heart to ascending aorta to arch of aorta to thoracic aorta to abdominal aorta to superior mesenteric artery to small intestine and half of large intestine to superior mesenteric vein to portal vein and also from abdominal aorta to inferior mesenteric artery to half of large intestine to inferior mesenteric vein to splenic vein to portal vein. From portal vein to liver to hepatic veins to inferior vena cava and return to heart.

956. From the abdominal aorta the ______________ arteries extend to the under surface of the diaphragm.

inferior phrenic

957. From the abdominal aorta the ______________ artery extends to the stomach, pancreas, duodenum, spleen, gallbladder, and liver.

celiac

958. From the abdominal aorta the ______________ artery extends to the small intestine and half of the large intestine.

superior mesenteric

959. From the abdominal aorta the ______________ artery extends to the other half of the large intestine.

inferior mesenteric

960. Label the indicated arteries in the abdominal aorta:

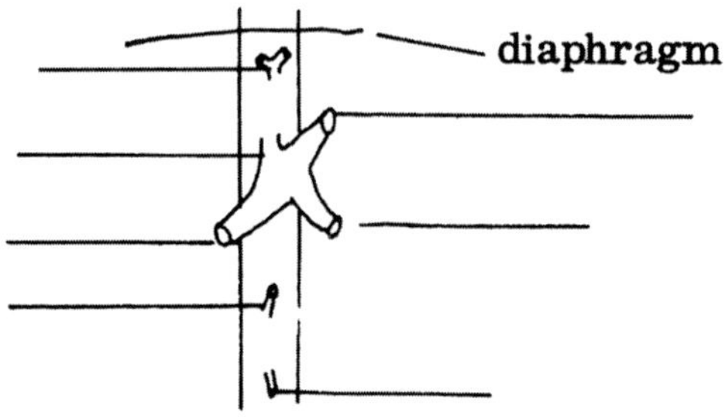

443. Label the following diagram:

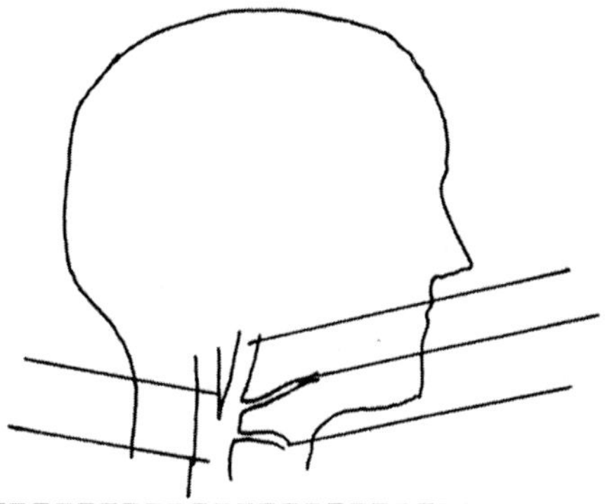

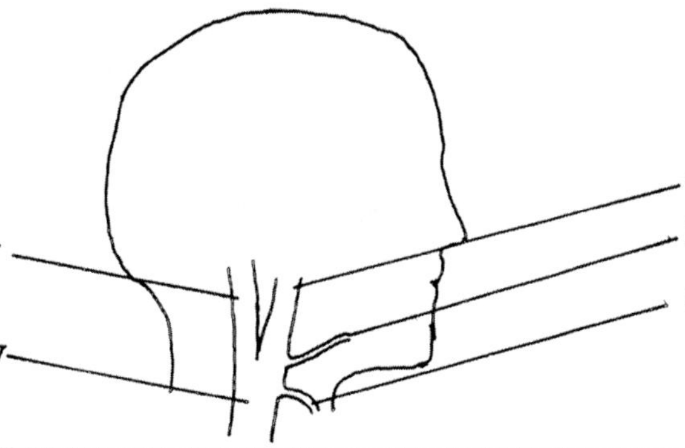

444. The next major branch of the external carotid artery is the external maxillary artery, the artery which leads towards the face.

445. The branch of the external carotid artery which leads towards the face is the ____________ maxillary artery.

external

446. The external ____________ artery is the branch of the external carotid artery which leads towards the face.

maxillary

447. The ____________ artery is the branch of the external carotid artery which leads towards the face.

external maxillary

448. The ____________ artery is the branch of the external carotid artery which leads towards the tongue.

lingual

449. The ____________ artery is the branch of the external carotid artery which leads towards the thyroid gland.

superior thyroid

936. The vein which receives the blood flowing from the inferior mesenteric vein is the ____________ vein.

---

splenic

---

937. The blood flows to the small intestine and half of the large intestine through the superior mesenteric artery and to the other half of the large intestine through the ____________ artery.

---

inferior mesenteric

---

938. From there it flows into the ________________ vein

---

inferior mesenteric

---

939. into the ______________ vein.

---

splenic

---

940. As before, the blood flowing from the digestive organs flows into the ______________ vein.

---

portal

---

941. Thus the blood flowing from the splenic vein empties into the ______________ vein,

---

portal

---

942. which carries it to the ________________.

---

liver

---

943. Time for a review. (If you don't need one, go on to frame 955).
The blood flows to the small intestine and half of the large intestine through the __________ artery.

---

superior mesenteric

---

944. From there it flows into the __________________ vein

---

superior mesenteric

---

945. and then into the ______________ vein

---

portal

---

946. which carries it to the ______________.

---

liver

---

947. From the liver the blood flows into the ________________ veins

---

hepatic

---

948. which empty into the ____________________.

---

inferior vena cava

---

949. The other half of the large intestine is supplied with blood by the ____________ artery

---

inferior mesenteric

---

950. which leads into the ________________ vein.

---

inferior mesenteric

---

450. Label the following diagram:

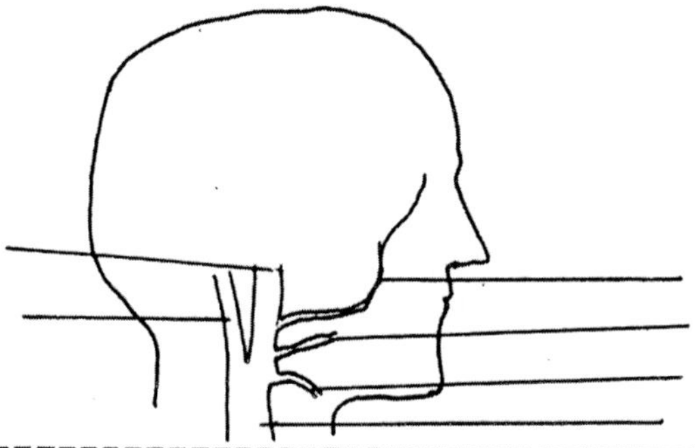

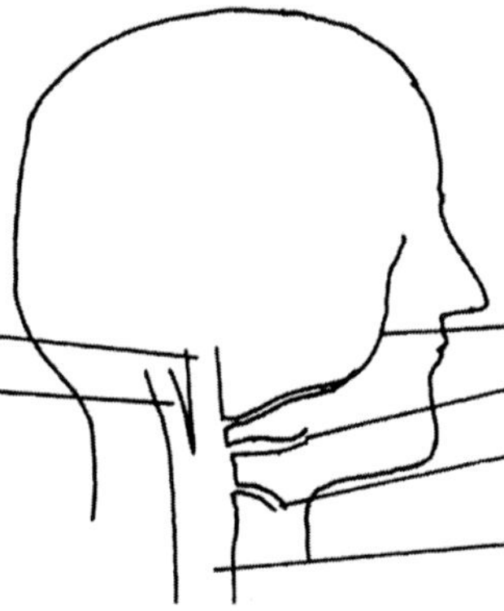

---

451. After the external maxillary artery, the next principal branch of the external carotid artery is the occipital artery, the artery which leads towards the scalp. (ok-sip'-i-t'l)

---

452. The branch of the external carotid artery which leads towards the scalp is the oc____________ artery.

occipital

---

453. The ____________ artery is the branch of the external carotid artery which extends towards the scalp.

occipital

---

454. The branch of the external carotid artery which leads towards the scalp is the ____________ artery.

occipital

---

455. The branch of the external carotid artery which extends towards the tongue is the ____________ artery.

lingual

---

456. The branch of the external carotid artery which extends towards the face is the ____________ artery.

external maxillary

---

457. The branch of the external carotid artery which leads towards the thyroid is the ____________ artery.

superior thyroid

927. The superior mesenteric artery leads to the small intestine and to half of the large intestine. The ____________ mesenteric artery leads to the other half of the large intestine.

---

inferior

---

928. The inferior ____________ artery extends to the other half of the large intestine.

---

mesenteric

---

929. The superior mesenteric artery supplies blood to the small intestine and half of the large intestine. The ________________ artery supplies blood to the other half of the large intestine.

---

inferior mesenteric

---

930. Label the parts of the abdominal aorta

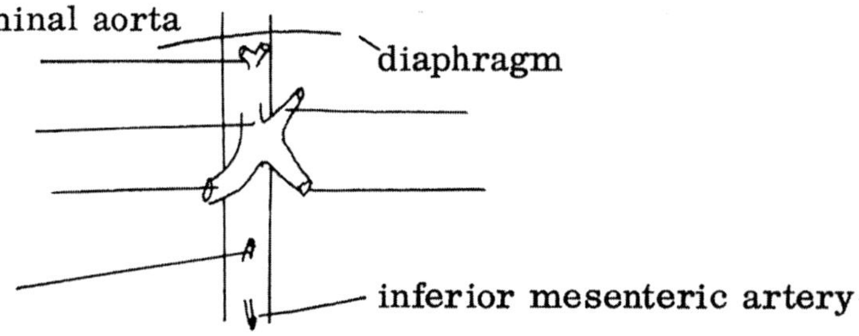

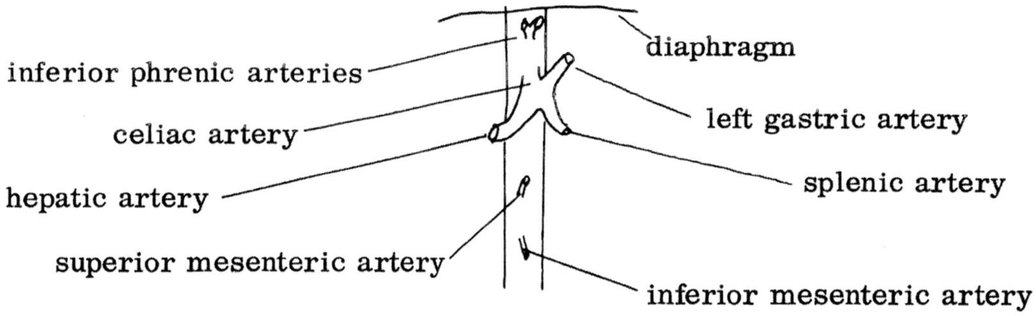

---

931. The ____________ artery extends to the small intestine and half of the large intestine.

---

superior mesenteric

---

932. The _____________ artery extends to the other half of the large intestine.

---

inferior mesenteric

---

933. The blood flows to the other half of the large intestine through the inferior mesenteric artery and from there through the corresponding vein, the ________________ vein.

---

inferior mesenteric

---

934. The vein receiving the blood flowing from the inferior mesenteric artery is the ________________ vein.

---

inferior mesenteric

---

935. From the inferior mesenteric vein the blood flows into the splenic vein.

---

458. The branch of the external carotid artery which extends towards the scalp is the ____________________ artery.

---

occipital

---

459. After the occipital branch the external carotid artery subdivides into two main branches, the superficial temporal artery and the internal maxillary artery.

---

460. The branch of the external carotid artery which extends upwards is the superficial temporal artery, the artery which leads towards the face and the scalp.

---

461. When the external carotid artery subdivides into two major branches, the one extending upwards is the ________________ temporal artery.

---

superficial

---

462. The upwards branch of the external carotid artery, the branch leading towards the face and scalp, is the superficial _______________ artery.

---

temporal

---

463. The ________________________ artery is the upwards branch of the external carotid artery which leads towards the face and scalp.

---

superficial temporal

---

464. Above the occipital artery the external carotid artery subdivides into two branches, the internal maxillary artery and the ______________________ artery.

---

superficial temporal

---

465. The external carotid artery subdivides into two branches, the superficial temporal artery and the internal maxillary artery. The internal maxillary artery leads towards the deep structures of the face, towards the covering of the brain, and towards the teeth and jaws.

---

466. The external carotid artery subdivides into two branches, the superficial temporal artery and the ________________ maxillary artery.

---

internal

---

467. The internal ______________ artery is the branch of the external carotid artery which leads towards the deep structures of the face and also towards the teeth and gums.

---

maxillary

---

468. The branch of the external carotid artery which leads towards the deep structures of the face, towards the covering of the brain, and towards the gums and teeth is the ________________ artery.

---

internal maxillary

---

469. Above the occipital artery the external carotid artery subdivides into two branches, the _________________________ artery and the _________________________ artery.

---

superficial temporal, internal maxillary

---

914. The vein receiving the blood from the intestines (and from the other digestive organs) is the ____________ vein.

portal

915. The portal vein carries the blood to the ____________.

liver

916. The blood flows from the abdominal aorta to the small intestine and to part of the large intestine through the ________________ artery.

superior mesenteric

917. The blood flows from the small intestine and part of the large intestine through the ________________ vein

superior mesenteric

918. which empties into the ____________ vein.

portal

919. The portal vein carries the blood to the ______________.

liver

920. From the liver the blood flows through the ______________ veins

hepatic

921. to empty into the ________________,

inferior vena cava

922. and back to the ________________ of the heart.

right atrium

923. The arteries extending from the abdominal aorta to the under surface of the diaphragm are the ________________ arteries.

inferior phrenic

924. The artery extending from the abdominal aorta to the stomach, pancreas, spleen, duodenum, and liver is the ____________ artery.

celiac

925. The artery extending from the abdominal aorta to the small intestine and part of the large intestine is the ____________________ artery.

superior mesenteric

926. The inferior mesenteric artery extends from the abdominal aorta to the other half of the large intestine.

470. Label the following diagram:

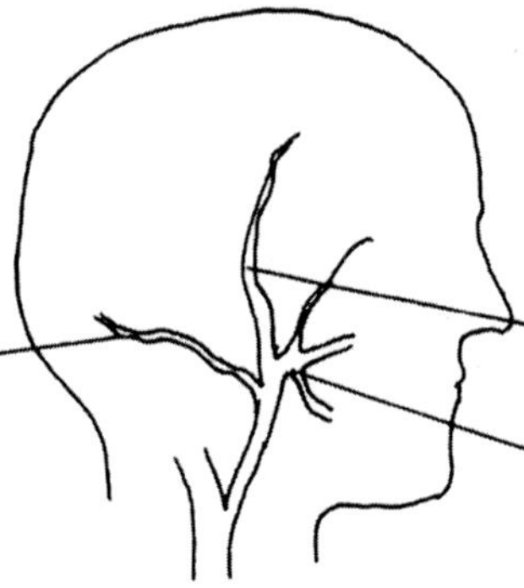

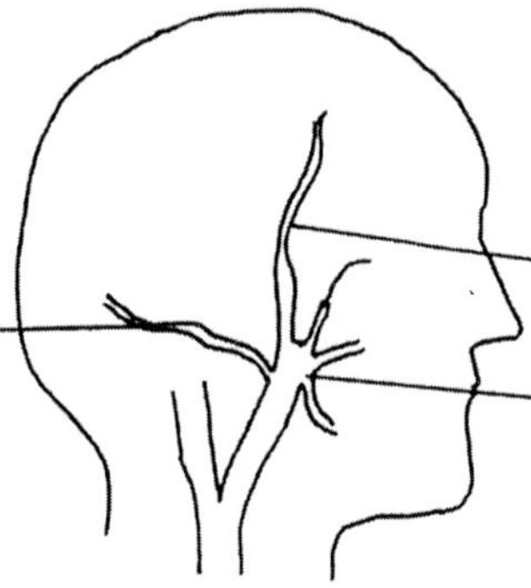

---

471. Time for a review. (If you are sure of the circulation thus far, proceed to frame 482). Blood from the heart flows through the ascending aorta, through the arch of the aorta, and then upwards through the right and left common ______________ arteries.

carotid

472. The common carotid arteries extend upward to the upper border of the thyroid cartilage where they subdivide into two branches, the ___________ artery and the ___________ artery.

external carotid, internal carotid

473. The first principal branch of the external carotid artery is the _____________ artery,

superior thyroid

474. the artery which leads towards the _____________________.

thyroid gland

475. The next branch of the external carotid artery is the ______________ artery,

lingual

476. the artery which extends towards the _____________.

tongue

477. After the lingual artery the next major branch of the external carotid artery is the _________________ artery, the artery which leads towards the face.

external maxillary

904. The superior mesenteric artery extends from the abdominal aorta to the ______________ and also to half of the large intestine.

small intestine

905. The artery extending from the abdominal aorta to the small and large intestines is the superior ______________ artery.

mesenteric

906. Label the following parts of the abdominal aorta:

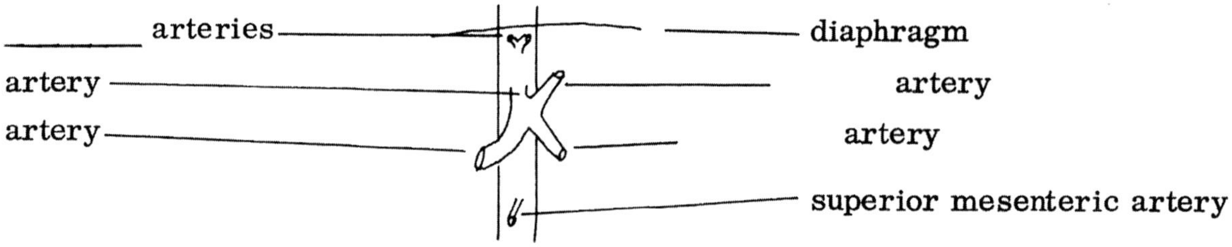

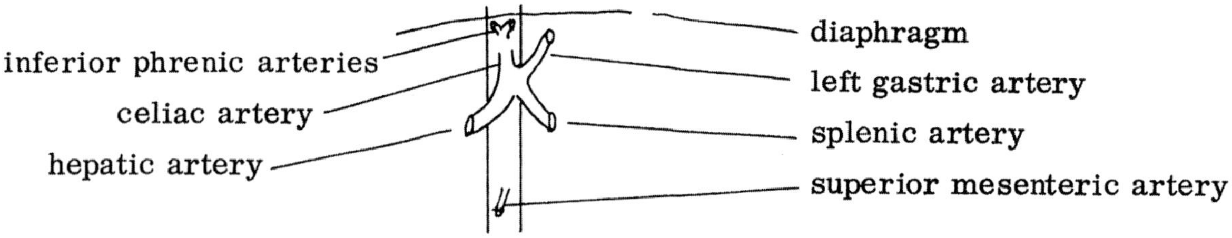

907. The blood flows to the small intestine and to part of the large intestine through the ______________ artery.

superior mesenteric

908. The blood flows to the small and large intestine through the superior mesenteric artery and from there through the corresponding vein, the ______________ vein.

superior mesenteric

909. The vein which receives the blood flowing from the small intestine and part of the large intestine is the superior ______________ vein.

mesenteric

910. Blood flows to the small and large intestine through the ______________ artery

superior mesenteric

911. and from the small and large intestine through the ______________ vein.

superior mesenteric

912. The small and large intestine are part of the digestive system and as before, the blood flowing from the digestive organs empties into the ______________ vein.

portal

913. The blood flowing from the small intestine and part of the large intestine flows into the superior mesenteric vein and then into the ______________ vein.

portal

478. After the external maxillary artery the next branch of the external carotid artery is the __________________ artery, leading towards the scalp.

occipital

479. From the occipital artery the external carotid artery subdivides into two branches, the ____________________ artery and the ______________________ artery.

superficial temporal, internal maxillary

480. The __________________________ artery extends upwards towards the face and scalp while the ______________________ artery extends towards the deep structures of the face, towards the covering of the brain and towards the teeth and jaws.

superficial temporal, internal maxillary

481. Label the following diagram:

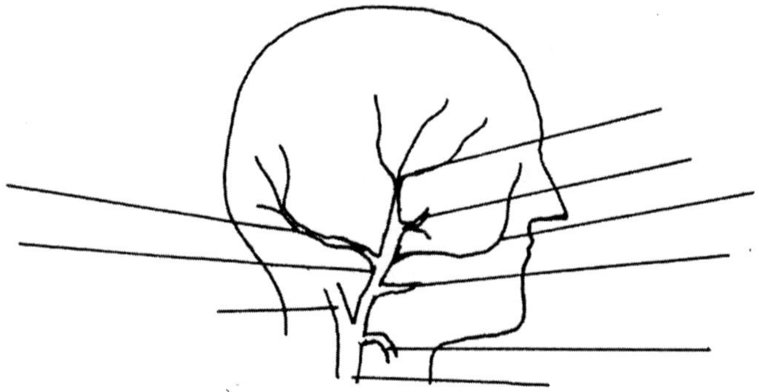

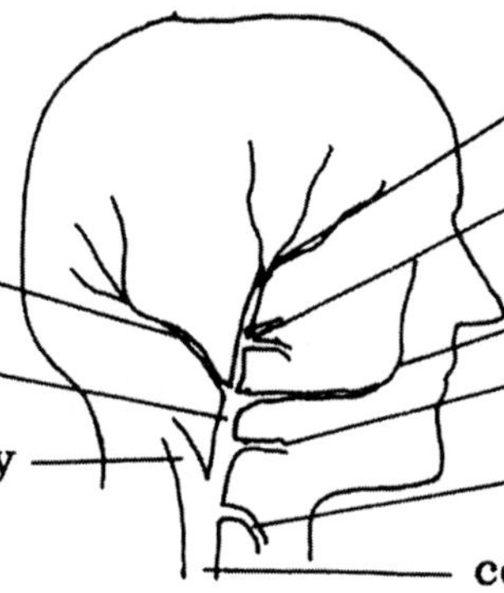

482. The blood flows through all of these arteries into arterioles, into capillaries, into venules and finally into two large veins, the posterior facial vein and the posterior auricular vein.

483. The two veins which receive blood from the face, scalp, and neck are the __________________ facial vein and the ___________________ auricular vein.

posterior, posterior

484. The two veins which receive blood from the head and neck regions are the posterior ___________________ vein and the posterior auricular vein.

facial

899. to the ______________ and then back to the heart.

---

inferior vena cava

---

900. Complete the following table:

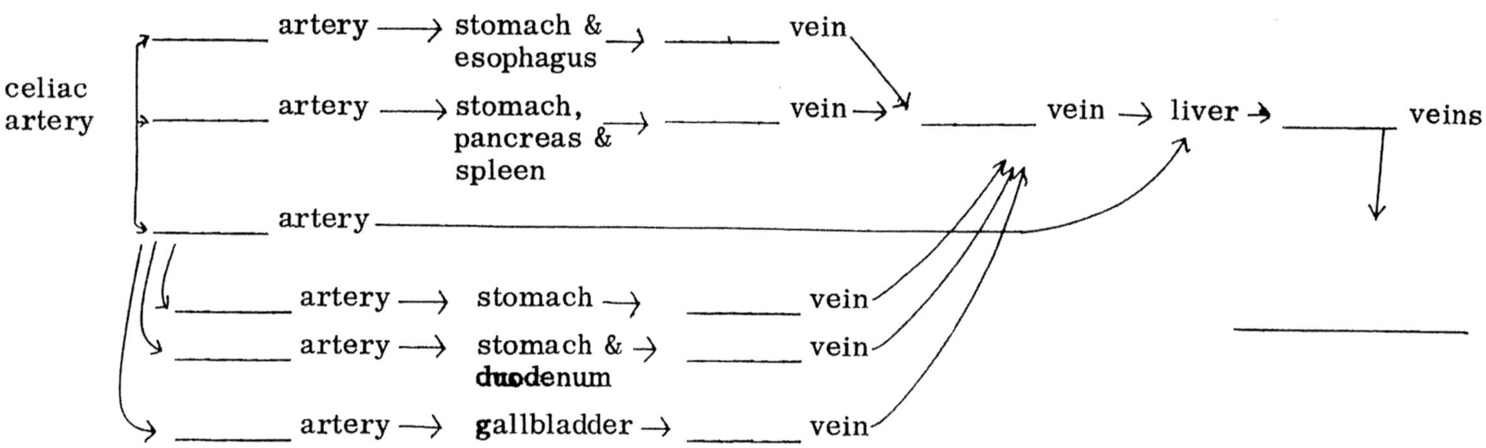

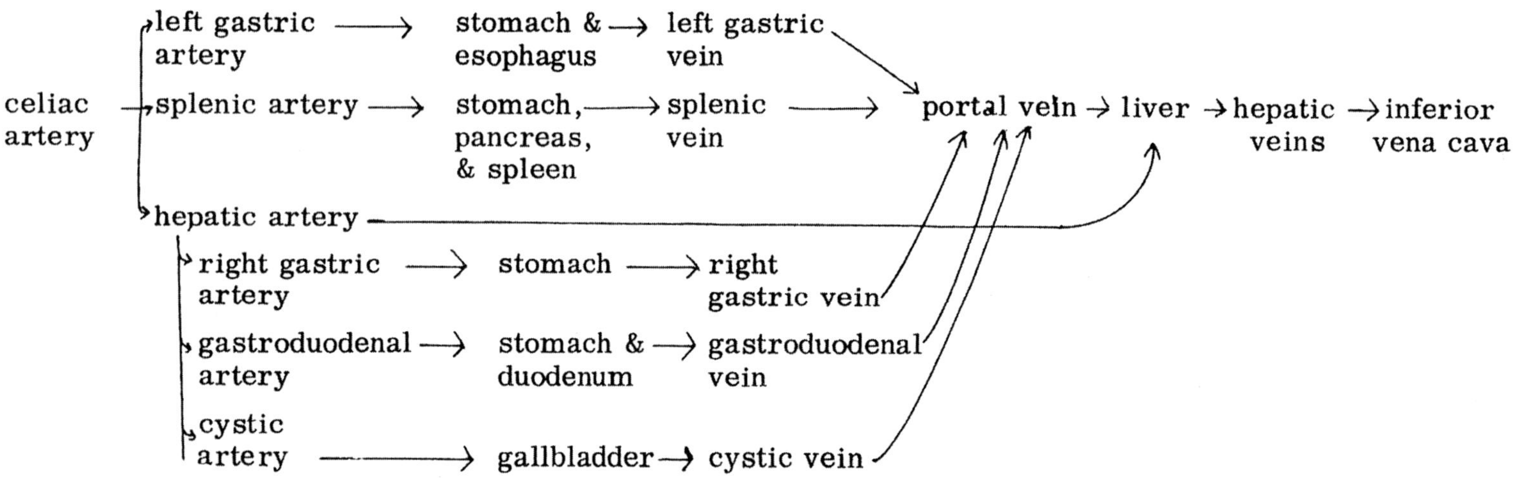

---

901. Trace the path of the blood from the right atrium to the stomach and return.

Right atrium through tricuspid valve to right ventricle through semilunar valves to pulmonary artery to lungs to pulmonary veins to left atrium through bicuspid valve to left ventricle through semilunar valves to ascending aorta to arch of aorta to thoracic aorta to abdominal aorta to celiac artery to left gastric artery, splenic artery, hepatic artery (to right gastric artery, gastroduodenal artery) to left gastric vein, splenic vein, right gastric vein, gastroduodenal vein to portal vein to liver to hepatic veins to inferior vena cava to right atrium.

---

902. The first branches of the abdominal aorta are the inferior phrenic arteries. The next branch is the celiac artery. After the celiac artery the next branch is the superior mesenteric artery, the artery leading to the small intestine and half of the large intestine.

---

903. The artery leading from the abdominal aorta to the small intestine and half of the large intestine is the ______________ mesenteric artery.

---

superior

---

485. Blood flows to the head and neck regions through the external carotid artery. It returns mainly through two veins, the posterior facial vein and the posterior ______________ vein.

---

auricular

---

486. The two veins which drain some of the blood from the head and neck are the posterior facial vein and the __________________________ vein.

---

posterior auricular

---

487. The blood to the head and neck flows from the branches of the external carotid artery into arterioles, capillaries, venules and then into the ________________ vein and the posterior auricular vein.

---

posterior facial

---

488. The two veins which receive the blood from the head and neck regions are the __________ and __________ veins.

---

posterior facial, posterior auricular

---

489. The posterior facial vein and the posterior auricular vein combine to form the external jugular vein.

---

490. The posterior facial vein and the posterior auricular vein on the right side of the head combine to form the right external jugular vein while those of the left side of the head form the ______________ external jugular vein.

---

left

---

491. The right and left external jugular veins are formed by the combining of the ____________ and ____________ veins.

---

posterior facial, posterior auricular

---

492. The posterior facial and the posterior auricular veins combine to form the right and left ________________ jugular veins.

---

external

---

493. The blood from the regions supplied by the external carotid artery flows into the posterior facial and the posterior auricular veins and then into the right and left external ____________ veins.

---

jugular

---

494. The right and left __________________ veins receive the blood flowing from the regions supplied by the external carotid arteries.

---

external jugular

---

495. The blood flows from the posterior facial and posterior auricular veins into the right and left _______________________ veins.

---

external jugular

---

496. The blood from the branches of the external carotid arteries flows into arterioles, capillaries, venules, and then into the ____________________ and ________________ veins.

---

posterior facial, posterior auricular

885. then into the ______________ vein

portal

886. to the ______________.

liver

887. From the liver the blood flows into the ______________ veins and then into the ______________.

hepatic, inferior vena cava

888. The ______________ artery sends a branch directly to the liver.

hepatic

889. The hepatic artery subdivides into three branches, the ______________ artery, the ______________ artery, and the ______________ artery.

right gastric, gastroduodenal, cystic

890. The ______________ artery extends from the hepatic artery to the stomach.

right gastric

891. The blood flowing to the stomach from the right gastric artery empties into the ______________ vein.

right gastric

892. The ______________ artery extends to the stomach and duodenum.

gastroduodenal

893. From the stomach and the duodenum the blood flows into the ______________ vein.

gastroduodenal

894. The ______________ artery extends to the gallbladder.

cystic

895. From the gallbladder the blood flows into the ______________ vein.

cystic

896. The blood flowing from the right gastric vein, the gastroduodenal vein, and the cystic vein empties into the ______________ vein,

portal

897. which carries the blood to the ______________.

liver

898. From the liver the blood flows into ______________ vein

hepatic

497. The posterior facial and the posterior auricular veins combine to form the ______________ vein on each side of the head.

---

external jugular

---

498. The external jugular veins empty into the subclavian veins, the right external jugular vein emptying into the right subclavian vein and the left external jugular vein emptying into the ______________ subclavian vein.

---

left

---

499. From the right and left external jugular veins the blood flows into the right and left ______________ veins.

---

subclavian

---

500. The right and left ______________ veins receive the blood flowing from the right and left external jugular veins.

---

subclavian

---

501 The posterior facial and posterior auricular veins combine to form the ______________ vein.

---

external jugular

---

502. The external jugular veins empty into the ______________ veins.

---

subclavian

---

503. The right and left subclavian veins empty into the right and left innominate veins.

---

504. The veins which receive the blood flowing from the right and left subclavian veins are the right and left ______________ veins.

---

innominate

---

505. The right and left innominate veins receive the blood flowing from the right and left ______________ veins.

---

subclavian

---

506 The right and left subclavian veins receive the blood flowing from the right and left ______________ veins.

---

external jugular

---

507. The external jugular veins receive the blood flowing from the ______________ and ______________ veins.

---

posterior facial, posterior auricular

---

508. The right and left innominate veins combine to form the superior vena cava which leads into the right atrium of the heart.

---

871. The ________________ receives the blood flowing from the hepatic veins and carries it back to the heart.

---

inferior vena cava

---

872. Time for another review. (If you don't want one, proceed to frame 900). The first branches of the abdominal aorta are the ____________ arteries which extend to the under surface of the diaphragm.

---

inferior phrenic

---

873. From the under surface of the diaphragm the blood flows into the ______________ veins,

---

inferior phrenic

---

874. and then empties into the ____________________.

---

inferior vena cava

---

875. The next branch of the abdominal aorta is the _____________ artery.

---

celiac

---

876. The three branches of the celiac artery are the _____________ artery, the _____________ artery, and the _____________ artery.

---

left gastric, splenic, hepatic

---

877. The __________________ artery extends to the stomach and esophagus.

---

left gastric

---

878. The blood flows to the stomach and esophagus through the left gastric artery and from there into the ________________ vein,

---

left gastric

---

879. which carries the blood to the ________________ vein

---

portal

---

880. from which it flows into the ________________.

---

liver

---

881. From the liver the blood flows into the ________________ veins

---

hepatic

---

882. to the ____________________ and back to the heart.

---

inferior vena cava

---

883. The _________________ artery extends to the pancreas, spleen, and stomach.

---

splenic

---

884. From there the blood flows into the ________________ vein

---

splenic

---

509. Label the following diagram:

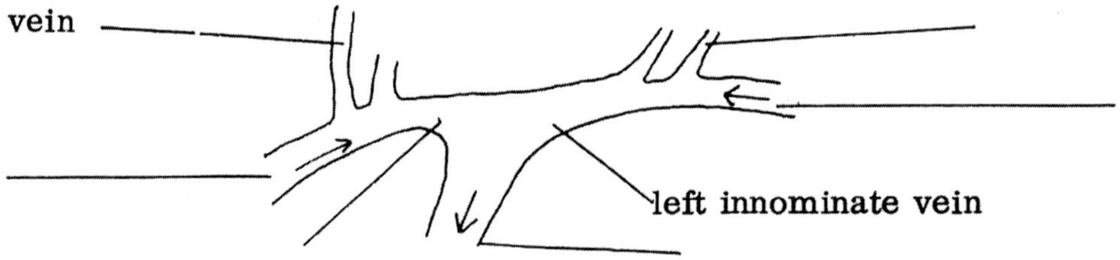

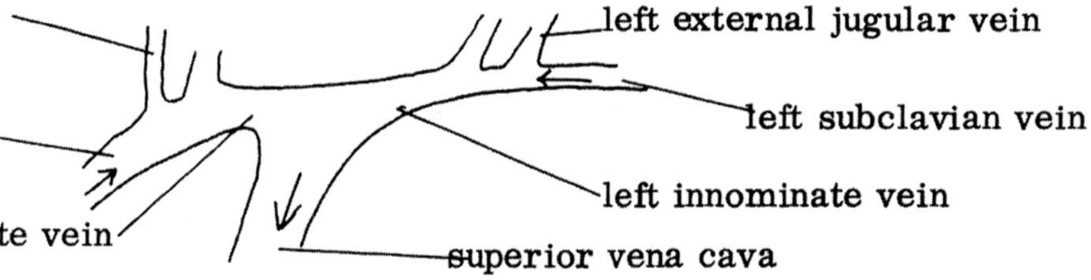

---

510. Review time again. (For those who know, frame 525 is the place to go).
The circulation from the heart to the left side of the head and neck is as follows:
The blood from the heart flows into the ______________ aorta

ascending

511. into the ______________ of the aorta

arch

512. into the __________________________ artery

left common carotid

513. into the ____________________ artery.

external carotid

514. The first branch of the external carotid artery is the ___________________ artery, which leads to the thyroid gland.

superior thyroid

515. The next branch, leading towards the tongue, is the _________________ artery.

lingual

516. The third principal branch of the external carotid artery is the ________________________ artery, leading towards the face.

external maxillary

517. The next branch, leading towards the scalp, is the _______________________ artery.

occipital

518. From the occipital artery the external carotid artery then subdivides into two branches, the ____________ artery and the ____________ artery.

superficial temporal, internal maxillary

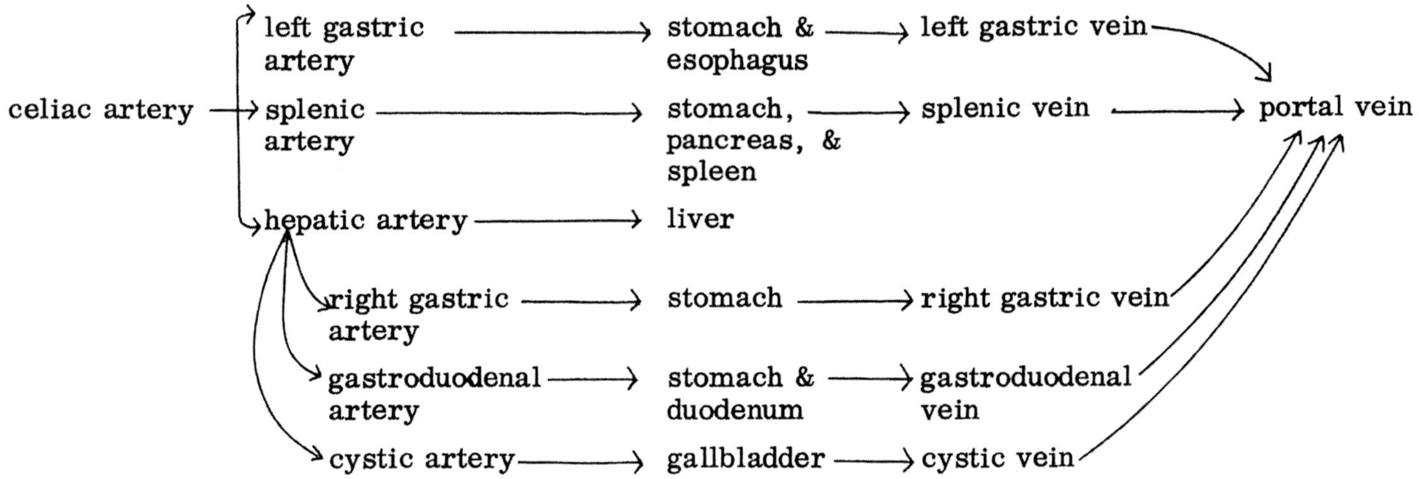

---

859. The portal vein carries the blood to the liver.

---

860. The blood from the digestive organs empties into the portal vein which carries it to the ___________.

liver

---

861. The liver receives the blood flowing from the ___________ vein.

portal

---

862. The blood flowing from the veins of the stomach, duodenum, pancreas, spleen, and gallbladder empties into the ___________ vein which carries it to the ___________.

portal, liver

---

863. From the liver the blood flows into the hepatic veins.

---

864. The veins which receive the blood flowing from the liver are the ___________ veins.

hepatic

---

865. The blood from the digestive organs flows into the ___________ vein

portal

---

866. which carries it to the ___________

liver

---

867. from which it flows through the ___________ veins.

hepatic

---

868. The veins which receive the blood flowing from the liver are the ___________ veins.

hepatic

---

869. From the hepatic veins the blood flows into the inferior vena cava and then back to the heart.

---

870. The blood flows from the liver through the ___________ veins and from there into the inferior ___________.

hepatic, vena cava

519. The blood from these branches of the external carotid artery flows into arterioles, capillaries, venules and then into two large veins, the __________ and __________ veins.

---

posterior facial, posterior auricular

520. The posterior facial and the posterior auricular veins combine to form the __________ vein. (Remember, this is the circulation to the left side of the head and neck).

---

left external jugular vein

521. The blood flows from the left external jugular vein into the __________ vein,

---

left subclavian

522. into the __________ vein

---

left innominate

523. into the __________

---

superior vena cava

524. and finally into the __________ of the heart.

---

right atrium

525. Label the following diagrams:

arterial flow

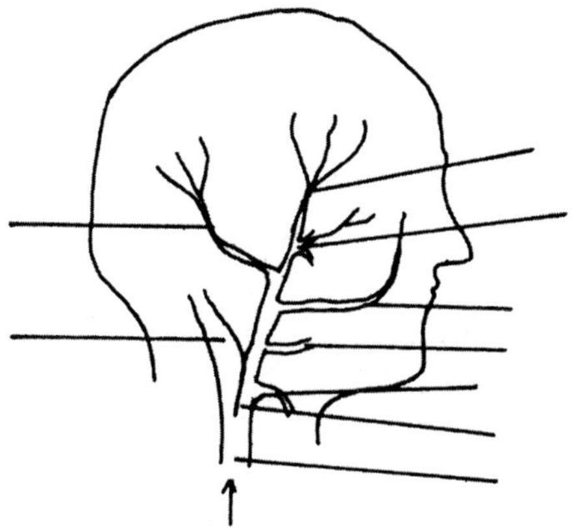

venous flow

---

850. The right ______________ artery extends to the stomach and from there the blood flows through the right ______________ vein.

gastric, gastric

851. The ______________ artery extends to the gallbladder.

cystic

852. From the gallbladder the blood flows through the ______________ vein.

cystic

853. The ____________________ artery supplies blood to the stomach and duodenum.

gastroduodenal

854. The blood flows to parts of the stomach and to the duodenum through the gastroduodenal artery and from there through the __________________ vein.

gastroduodenal

855. All of the blood coming from the celiac artery (except for the branch from the hepatic artery direct to the liver) flows to the indicated parts of the digestive system through the proper arteries and from there through the corresponding veins and then empties into the portal vein.

856. The vein which receives the blood flowing from the digestive organs is the __________ vein.

portal

857. The blood flows to the digestive organs through the arteries and from there through the corresponding veins and then into the ______________ vein.

portal

858. Complete the following table:

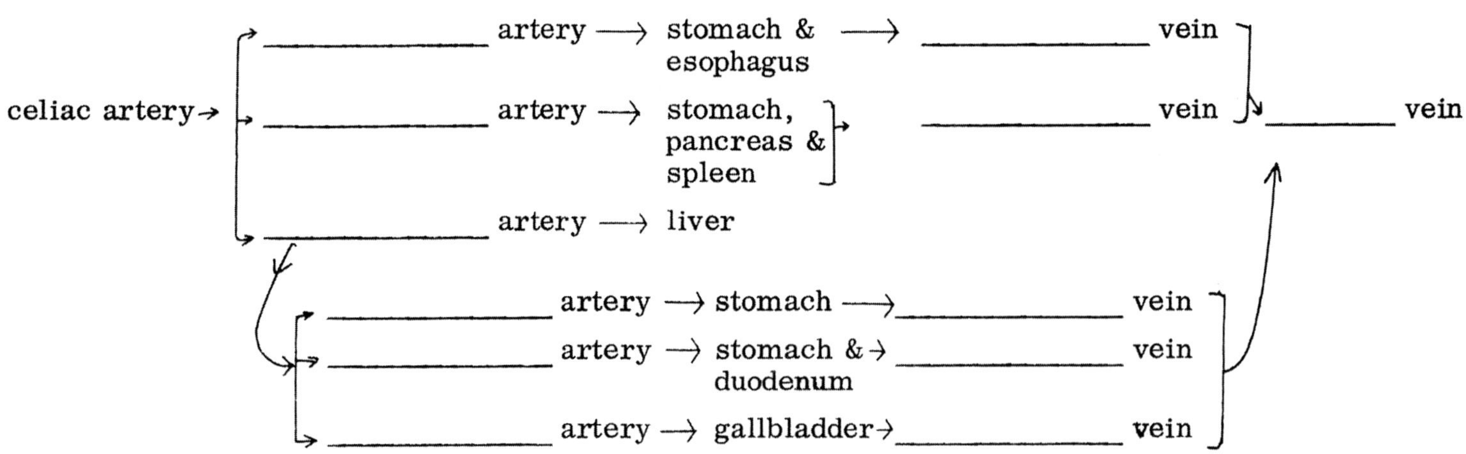

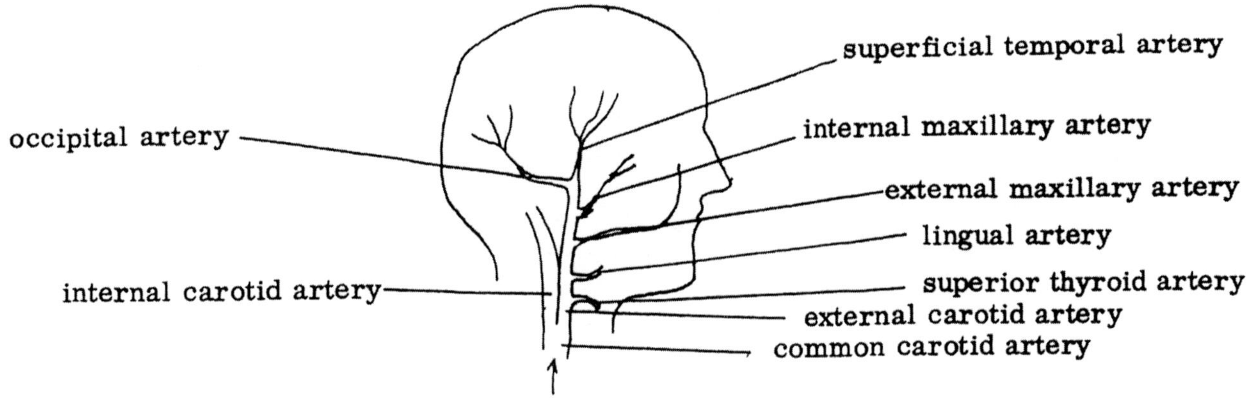

---

526. Describe the circulation of the blood from the heart to the <u>right</u> side of the head and return.

Heart, ascending aorta, arch of the aorta, innominate artery, right common carotid artery, external carotid artery, (branches, superior thyroid artery, lingual artery, external maxillary artery, occipital artery, internal maxillary artery and superficial temporal artery) arterioles, capillaries, venules, posterior facial vein and posterior auricular vein, right external jugular vein, right subclavian vein, right innominate vein, superior vena cava, right atrium of the heart.

---

838. The three branches of the hepatic artery are the ______________ artery, the ______________ artery, and the ______________ artery.

---

right gastric, gastroduodenal, cystic

---

839. The hepatic artery supplies blood directly to the liver. It also subdivides into three branches,
the ______________ artery extending to the ______________,
the ______________ artery extending to the ______________, and
the ______________ artery extending to the ______________.

---

right gastric, stomach
gastroduodenal, stomach & duodenum
cystic, gallbladder

---

840. Review time. The first branches of the abdominal aorta are the ______________ arteries which lead to the under surface of the diaphragm.

---

inferior phrenic

---

841. From the under surface of the diaphragm the blood flows through the ____________ veins.

---

inferior phrenic

---

842. The next branch of the abdominal aorta is the ______________ artery which subdivides into three branches.

---

celiac

---

843. The three branches of the celiac artery are the ______________ artery, the ______________ artery, and the ______________ artery.

---

left gastric, splenic, hepatic

---

844. The _____________ artery extends to parts of the stomach and esophagus.

---

left gastric

---

845. The blood flows from parts of the stomach and esophagus supplied by the left gastric artery through the ___________________ vein.

---

left gastric

---

846. The ________________ artery supplies blood to the pancreas, stomach and spleen.

---

splenic

---

847. The blood flows to parts of the stomach, the pancreas and the spleen through the splenic artery and from there through the _______________ vein.

---

splenic

---

848. The _________________ artery extends one branch directly to the liver.

---

hepatic

---

849. The hepatic artery also divides into three branches, the ______________ artery, the ______________ artery, and the ______________ artery.

---

right gastric, gastroduodenal, cystic

VI. THE BRAIN

527. The brain is supplied with blood by means of the common carotid arteries and the vertebral arteries.

---

528. The right common carotid artery arises from the innominate artery and the left common carotid artery from the arch of the aorta. The right and left vertebral arteries arise from the corresponding subclavian arteries.

---

529. Label the following diagram:

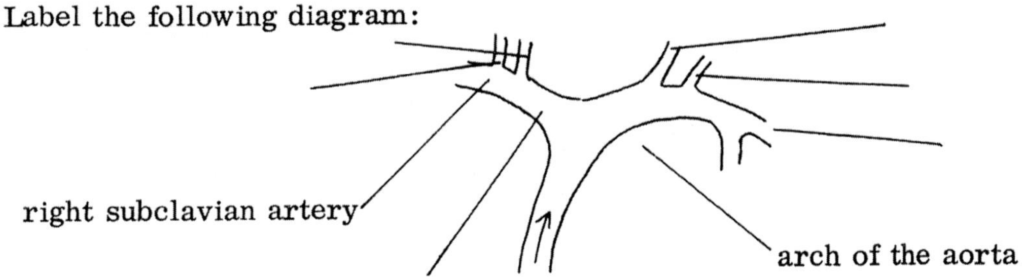

- -

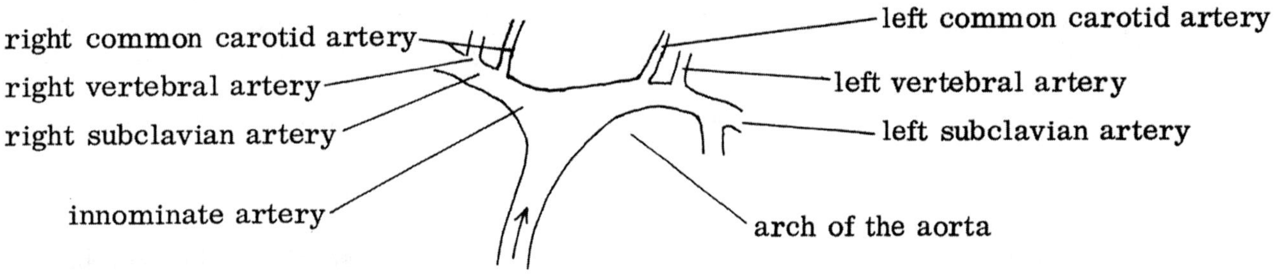

---

530. The right and left vertebral arteries extend upwards to the base of the brain where they unite to form the basilar artery.

---

531. When the right and left vertebral arteries combine at the base of the brain, they form the ________________ artery.

- -

basilar

---

532. The _____________ artery is formed by the union of the right and left vertebral arteries.

- -

basilar

---

533. The right and left __________________ arteries unite at the base of the brain to form the basilar artery.

- -

vertebral

---

534. The right vertebral artery arises from the right ________________ artery and the left vertebral artery from the ____________________ artery.

- -

subclavian, left subclavian

---

829. Complete the following table:

celiac artery	______________ artery →	parts of stomach and esophagus →	______________ vein	
	______________ artery →	pancreas, spleen and parts of stomach →	______________ vein	
	______________ artery			

---

celiac artery	left gastric artery →	parts of stomach and esophagus →	left gastric vein
	splenic artery →	pancreas, spleen and parts of stomach →	splenic vein
	hepatic artery		

---

830. The hepatic artery supplies blood directly to the liver. It also gives off three branches, the right gastric artery, the gastroduodenal artery, and the cystic artery.

---

831. The hepatic artery divides into the right gastric artery, the gastroduodenal artery, and the cystic artery. It also supplies blood directly to the ______________.

liver

---

832. The hepatic artery supplies blood directly to the liver. It also gives off three branches, the right gastric artery which extends to the ____________, the gastroduodenal artery which extends to the __________ and __________ and the cystic artery which extends to the gall bladder.

stomach, stomach & duodenum

---

833. The three branches of the hepatic artery are the right __________ artery, the gastro-duodenal artery, and the cystic artery.

gastric

---

834. The right __________ artery extends from the hepatic artery to the stomach.

gastric

---

835. The ______________ artery extends from the hepatic artery to the stomach and the duodenum.

gastroduodenal

---

836. The hepatic artery subdivides into the ____________ artery, the ____________ artery, and the cystic artery.

right gastric, gastroduodenal

---

837. The __________ artery extends to the gall bladder.

cystic

535. Label the following diagram:

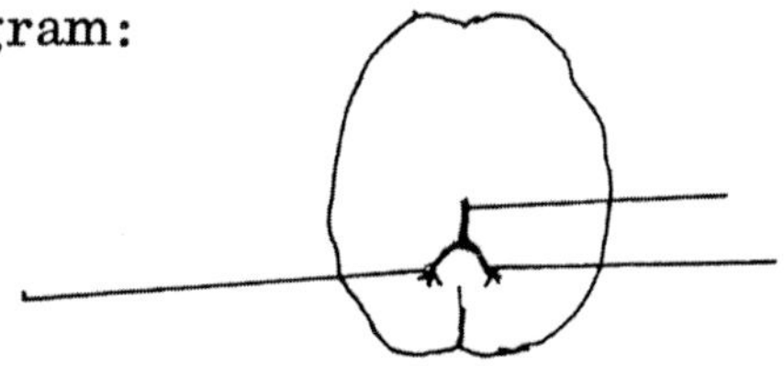

base of the brain

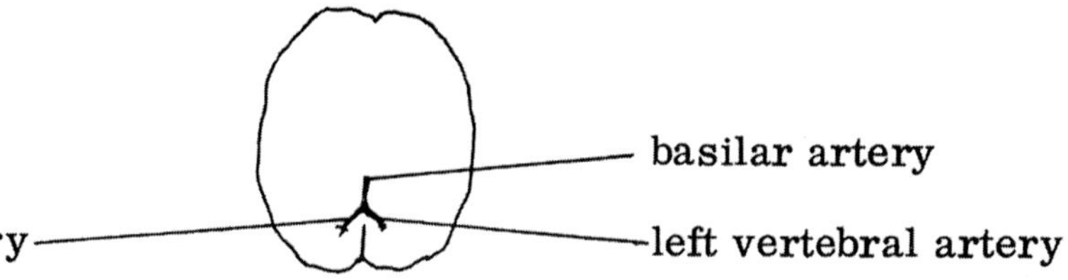

base of the brain

---

536. The basilar artery branches into the right and left posterior cerebral arteries.

---

537. The two branches of the basilar artery are the right and the left ____________ cerebral arteries.

posterior

---

538. The basilar artery subdivides into two branches, the right and left posterior ____________ arteries.

cerebral

---

539. The right and left ____________ arteries are the two branches of the basilar artery.

posterior cerebral

---

540. The basilar artery branches into the ________________________ arteries.

right and left posterior cerebral

---

541. The right and left vertebral arteries unite at the base of the brain to form the ____________ artery.

basilar

---

542. The basilar artery branches into the ______________________________ arteries.

right and left posterior cerebral

---

543. Label the following diagram:

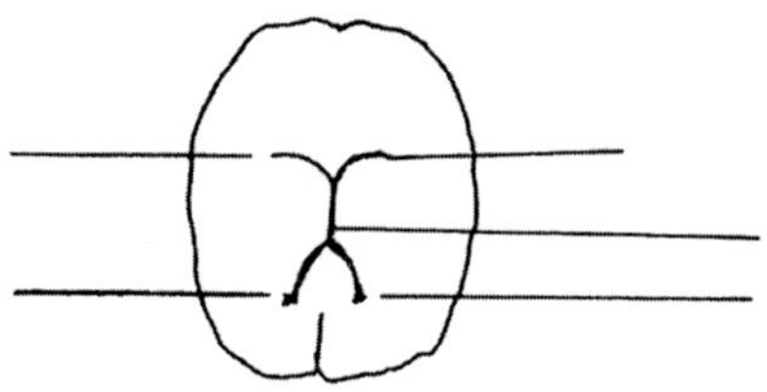

base of brain

---

820. The celiac artery divides into the ______________ artery, the ______________ artery, and the ______________ artery.

---

left gastric, hepatic, splenic

821. Label:

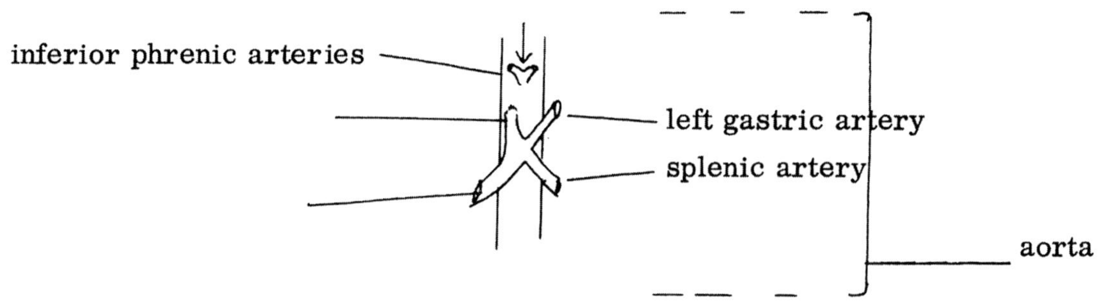

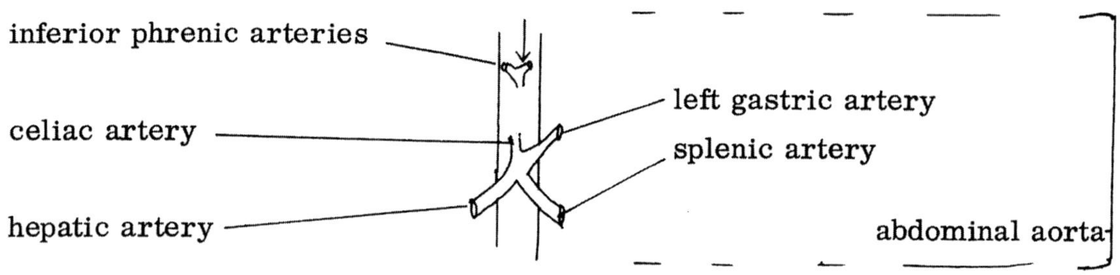

822. The first branches of the abdominal aorta are the ______________ arteries which lead to the under surface of the diaphragm.

---

inferior phrenic

823. The next branch of the abdominal aorta is the ______________ artery which divides into the ______________ artery, the ______________ artery, and the ______________ artery.

---

celiac, left gastric, hepatic, splenic

824. The arteries extending from the celiac artery go to certain organs and from these organs the blood flows through the corresponding veins.

825. The left gastric artery leads to parts of the stomach and esophagus.

826. The blood flows to the parts of the stomach and esophagus through the left gastric artery and from there through the ________________ vein.

---

left gastric

827. The splenic artery extends to the pancreas, to the spleen and to parts of the stomach.

828. The blood flows to the pancreas, spleen and parts of the stomach through the splenic artery and from there through the ______________ vein.

---

splenic

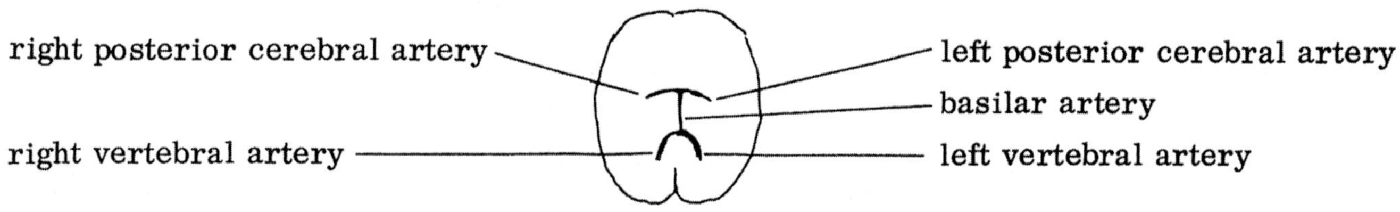

---

544. The right and left common carotid arteries extend upwards along the trachea until they reach the upper border of the thyroid cartilage where they subdivide into the external carotid artery and the internal carotid artery.

---

545. The external carotid arteries supply blood to the head and neck while the internal carotid arteries supply the brain and also the eyes and nose.

---

546. The right and left common carotid arteries subdivide into the external carotid artery and the ________________ artery.

internal carotid

---

547. The first major branch of the internal carotid artery is the ophthalmic artery, the artery carrying blood to the eye.

---

548. The ophthalmic artery is the branch of the internal carotid artery which supplies blood to the ____________.

eye

---

549. The branch of the internal carotid artery which supplies blood to the eye is the ____________ artery.

ophthalmic

---

550. The first principal branch of the internal carotid artery is the ________________ artery, the artery which supplies blood to the eye.

ophthalmic

---

551. The first principal branch of the internal carotid artery is the ophthalmic artery. The other branches go to the brain.

---

552. When the right and left internal carotid arteries reach the base of the brain they divide into several branches. The first branches are the posterior communicating arteries (right and left).

---

553. At the base of the brain the first branch of the internal carotid artery is the ______________ communicating artery.

posterior

---

554. When the right and left internal carotid arteries reach the base of the brain, their first branch is the posterior ________________ arteries.

communicating

---

555. The ___________________ arteries form the first branch of the internal carotid arteries at the base of the brain.

posterior communicating

807. The blood flowing to the upper surface of the diaphragm flows from the thoracic aorta to the _________________ arteries

superior phrenic

808. to the upper surface of the diaphragm and then to the _____________________ veins

superior phrenic

809. to the _____________________ and back to the heart.

superior vena cava

810. After the inferior phrenic arteries, the next artery extending from the abdominal aorta is the celiac artery which subdivides into three branches.

811. The three branches of the celiac artery are the left gastric artery, the hepatic artery, and the splenic artery.

812. The celiac artery subdivides into three branches, the ______________ gastric artery, the hepatic artery, and the splenic artery.

left

813. The left gastric artery, the hepatic artery, and the splenic artery are the three branches of the _____________ artery.

celiac

814. One of the three branches of the celiac artery is the left _______________ artery.

gastric

815. The three branches of the celiac artery are the hepatic artery, the splenic artery, and the ______________ artery.

left gastric

816. The celiac artery divides into three branches, the left gastric artery, the _______________ artery, and the splenic artery.

hepatic

817. In addition to the splenic artery, the other branches of the celiac artery are the ___________ artery and the _____________ artery.

left gastric, hepatic

818. The celiac artery divides into the left gastric artery, the hepatic artery, and the ___________ artery.

splenic

819. The left gastric artery, the hepatic artery, and the splenic artery are the three branches of the ______________ artery which extends from the _______________ aorta.

celiac, abdominal

556 The right and left common carotid arteries subdivide into the ________________________ arteries.

---

external and internal carotid

557. The first principal branch of the internal carotid artery is the __________________ artery which leads to the eye.

---

ophthalmic

558. At the base of the brain the first branches of the internal carotid arteries are the ____________________ arteries.

---

posterior communicating

559. The posterior communicating arteries lead to the corresponding posterior cerebral arteries.

560. The posterior communicating arteries "communicate" with the posterior _______________ arteries.

---

cerebral

561. At the base of the brain the first branch of the internal carotid artery is the posterior communicating artery which connects with the __________________ cerebral artery.

---

posterior

562. The posterior communicating arteries connect with the _______________________ arteries.

---

posterior cerebral

563. Reviewing: The right and left vertebral arteries extend upwards to the base of the brain where they unite to form the ________________ artery.

---

basilar

564. The basilar artery subdivides into the right and left __________________ arteries.

---

posterior cerebral

565. The posterior cerebral arteries are connected to the internal carotid arteries by the __________________ arteries.

---

posterior communicating

794. The first group of arteries extending from the abdominal aorta are the inferior phrenic arteries, the arteries leading to the under surface of the diaphragm.

---

795. The arteries leading to the under surface of the diaphragm are the ____________ phrenic arteries.

inferior

796. The inferior _____________ arteries lead to the under surface of the diaphragm.

phrenic

797. The arteries extending from the abdominal aorta to the under surface of the diaphragm are the ________________ arteries.

inferior phrenic

798. The arteries extending from the thoracic aorta to the upper surface of the diaphragm are the _________________ arteries.

superior phrenic

799. The blood flows to the upper surface of the diaphragm through the superior phrenic arteries and from there through the superior phrenic veins. Likewise the blood flows to the under surface of the diaphragm through the inferior phrenic arteries and from there through the __________ veins.

inferior phrenic

800. The veins which receive the blood flowing from the under surface of the diaphragm are the _____________________ veins.

inferior phrenic

801. The blood flows from the inferior phrenic veins to the inferior vena cava.

---

802. From the under surface of the diaphragm the blood flows into the inferior phrenic veins and then into the inferior ________________.

vena cava

803. The blood from the superior phrenic veins flows into the superior vena cava while that from the inferior phrenic veins flows into the ________________________.

inferior vena cava

804. The blood flowing to the under surface of the diaphragm flows from the abdominal aorta to the ________________ arteries

inferior phrenic

805. to the lower surface of the diaphragm and then to the _________________ veins

inferior phrenic

806. to the ________________ and back to the heart.

inferior vena cava

566. Label the following diagram:

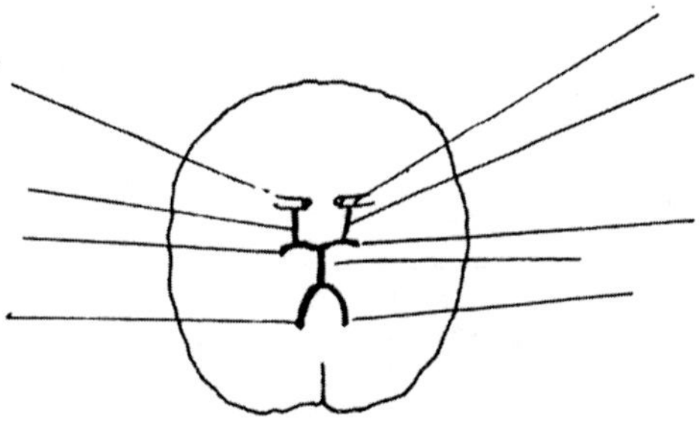

base of brain

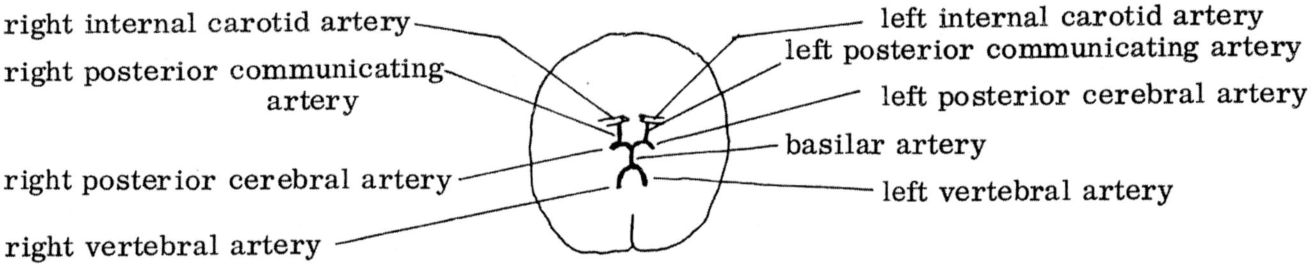

base of brain

---

567. After giving off its first branch, the internal carotid artery subdivides into two parts, the anterior cerebral artery and the middle cerebral artery.

---

568. When the internal carotid arteries reach the base of the brain they give off as their first branch the posterior communicating arteries and then divide into two branches, the ____________ cerebral arteries and the middle cerebral arteries.

anterior

---

569. The anterior and ______________ cerebral arteries are the two terminal branches of the internal carotid artery at the base of the brain.

middle

---

570. When the internal carotid arteries subdivide they form the anterior and middle ______________________ arteries.

cerebral

---

571. At the base of the brain the internal carotid arteries have as their first branch the ________________________ arteries.

posterior communicating

---

572. They then subdivide into the ___________________ artery and the _______________ artery.

anterior cerebral, middle cerebral

Artery	Body Part	Vein		
pericardial →	pericardium →	pericardial	left innominate vein	superior vena cava
superior phrenic →	upper surface of diaphragm →	superior phrenic	left innominate vein	superior vena cava
esophageal →	esophagus →	esophageal	left innominate vein	superior vena cava
bronchial →	bronchi →	bronchial	azygos vein	superior vena cava
intercostal →	muscles & skin of chest wall →	intercostal	azygos vein	superior vena cava

---

789. Trace the flow of blood from the heart to the esophagus and return.

From heart to ascending aorta to arch of aorta to thoracic aorta to esophageal arteries to esophagus to esophageal veins to left innominate vein to superior vena cava to heart.

---

790. Trace the path of blood flow from the heart to the upper surface of the diaphragm and return.

From the heart through the ascending aorta to arch of aorta to thoracic aorta to superior phrenic arteries to upper surface of the diaphragm to superior phrenic veins to left innominate vein to superior vena cava to heart.

---

791. Trace the flow of blood from the heart to the muscles and skin of the chest wall and return.

From the heart through the ascending aorta, to arch of aorta to thoracic aorta to intercostal arteries to muscles and skin of chest wall to intercostal veins to azygos vein to superior vena cava to heart.

VIII. THE ABDOMINAL AORTA

792. As the blood leaves the heart it flows into the ascending aorta and then the arch of the aorta. From the arch of the aorta it flows into the thoracic aorta which extends down to the diaphragm. Below the diaphragm the blood enters the abdominal aorta.

---

793. Label the following diagram:

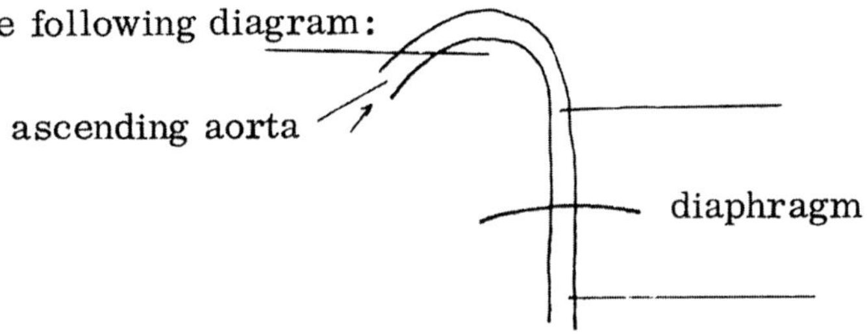

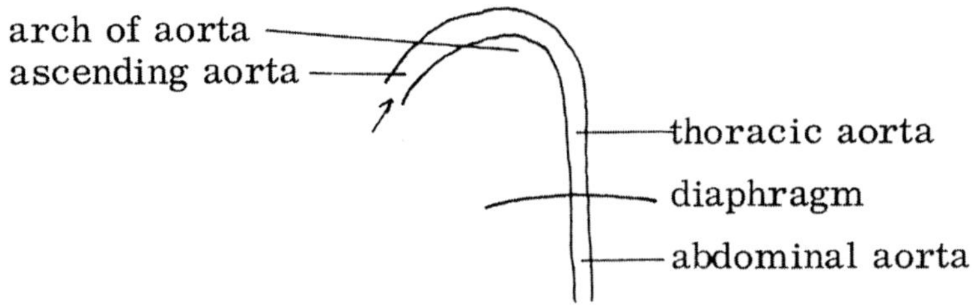

---

573. Label the following diagram:

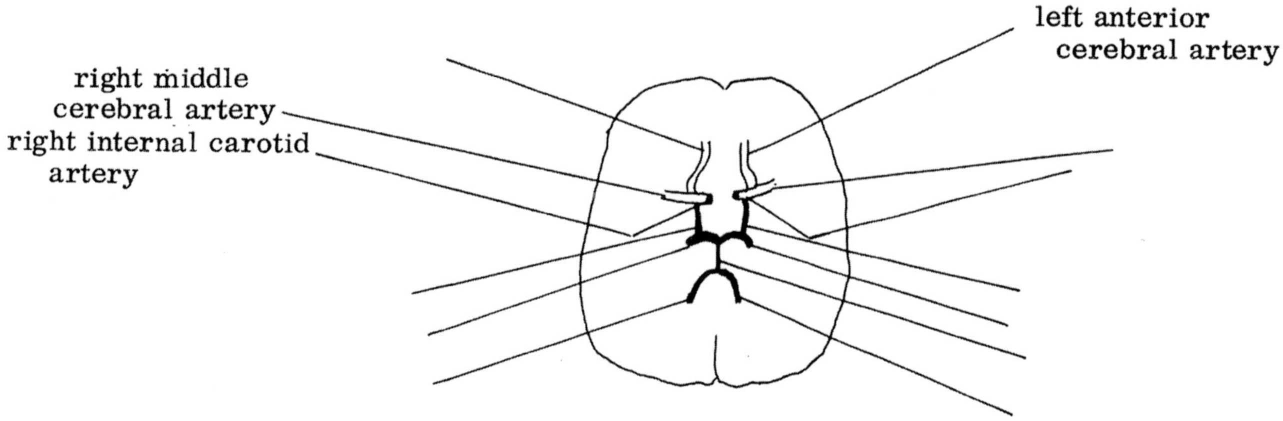

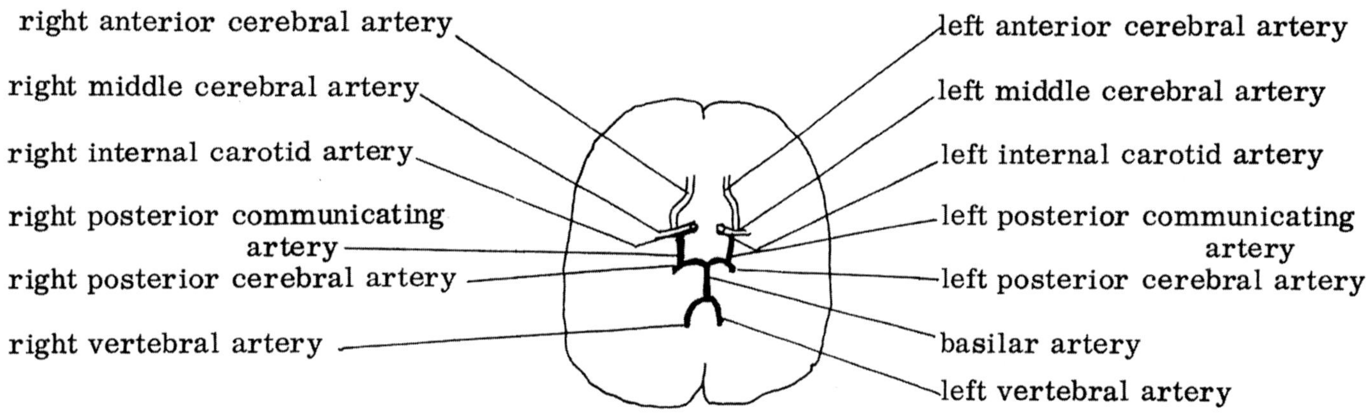

574. The right and left anterior cerebral arteries are connected by means of the anterior communicating artery.

575. The artery which connects the right and left anterior cerebral arteries is the ____________ communicating artery.

anterior

576. The anterior ____________ artery connects the right and left anterior cerebral arteries.

communicating

577. The artery which connects the right and left anterior cerebral arteries is the ____________ artery.

anterior communicating

578. At the base of the brain the right and left internal carotid arteries give off branches, the ____________ arteries

posterior communicating

579. and then subdivide into the ____________ arteries and the ____________ arteries.

middle cerebral, anterior cerebral

776. From the esophagus the blood flows into the ______________ veins

esophageal

777. into the ________________

left innominate vein

778. into the ______________.

superior vena cava

779. The parietal branches of the thoracic aorta are the ______________ arteries and the ______________ arteries.

intercostal, superior phrenic

780. The intercostal arteries extend to the ______________________.

muscles and skin of the chest wall

781. The superior phrenic arteries extend to the _________________________.

upper surface of the diaphragm

782. From the muscles and skin of the chest wall the blood flows into the ______________ veins

intercostal

783. into the ____________ vein

azygos

784. into the ___________.

superior vena cava

785. From the upper surface of the diaphragm the blood flows through the ______________ veins

superior phrenic

786. into the ________________ vein

left innominate

787. and then into the ________________.

superior vena cava

788. Complete the following table:

Artery		Body part		vein				
pericardial	→	________	→	________	⎤			
superior phrenic	→	________	→	________	⎬ →	________	⎤	
esophageal	→	________	→	________	⎦		⎬ →	________
bronchial	→	________	→	________	⎤			
intercostal	→	________	→	________	⎦ →	________	⎦	

580. The right and left anterior cerebral arteries are connected by means of the ___________ artery.

---

anterior communicating

---

581. Label the following diagram:

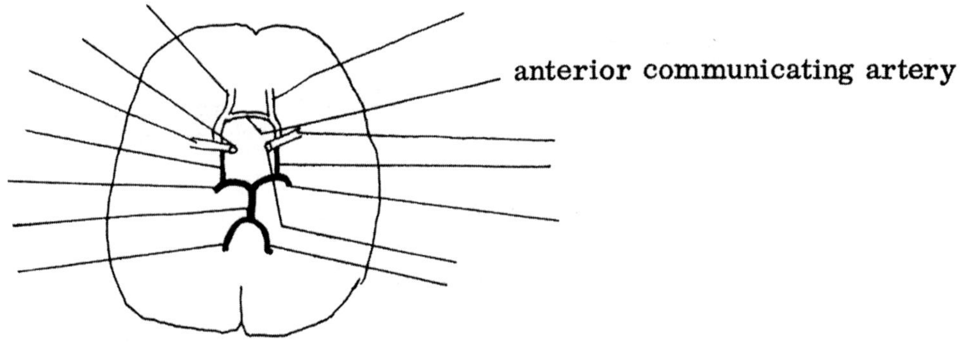

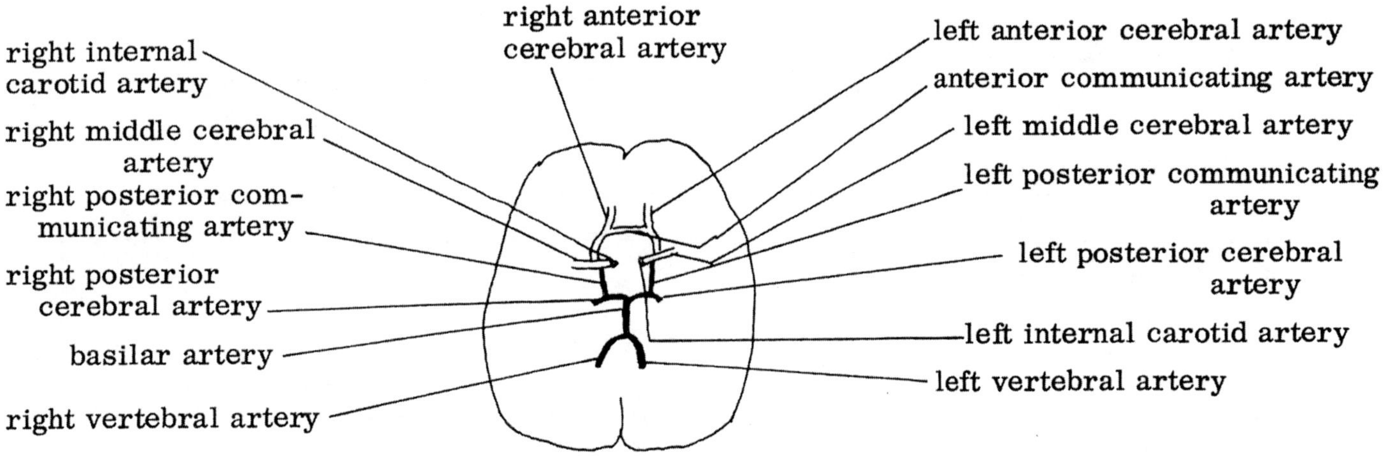

---

582. The above diagram represents the "Circle of Willis", an arterial circle at the base of the brain.

---

583. The Circle of Willis is formed by the anterior cerebral arteries, which are branches of the ________________ arteries

---

internal carotid

---

584. and the posterior cerebral arteries which are branches of the ______________ artery.

---

basilar

---

585. The anterior cerebral arteries are connected by means of the __________________ artery.

---

anterior communicating

---

761. and from the bronchi through the ______________ veins.

bronchial

762. From the bronchial veins the blood flows into the ____________ vein

azygos

763. into the _______________

superior vena cava

764. and then to the __________________ of the heart.

right atrium

765. Time for a review. (If you know, then on to frame 788 you go).
The thoracic aorta has two types of branches, the ___________ arteries and the ___________ arteries.

visceral, parietal

766. The visceral arteries extending from the thoracic aorta are the _____________, the _____________, and the _____________ arteries.

pericardial, esophageal, bronchial

767. The pericardial arteries extend to the __________________.

pericardium

768. The bronchial arteries extend to the _______________

bronchi

769. and the esophageal arteries to the __________________.

esophagus

770. From the pericardium the blood flows into the ______________ veins

pericardial

771. and then into the __________________ vein

left innominate

772. to the ________________.

superior vena cava

773. From the bronchi the blood flows into the _________________ veins

bronchial

774. into the _______________ vein

left innominate

775. and into the ________________.

superior vena cava

586. The posterior cerebral arteries and the internal carotid arteries are connected by the ________________ arteries.

posterior communicating

587. Time for another review. (If you are sure you know, then frame 599 is the place for you to go).
The right and left vertebral arteries arise from the corresponding ____________ arteries.

subclavian

588. The right and left vertebral arteries extend to the base of the brain where they unite to form the _____________ artery.

basilar

589. The basilar artery subdivides into the _________________ arteries.

right and left posterior cerebral

590. The right common carotid artery arises from the ____________ artery and the left common carotid artery from the arch of the aorta.

innominate

591. The right and left common carotid arteries rise into the neck where they subdivide into the _____________________ arteries.

internal and external carotid

592. The internal carotid artery gives off one major branch, the ____________ artery which extends to the eye.

ophthalmic

593. The internal carotid arteries then extend to the base of the brain where they each give off a branch, the _________________ arteries.

posterior communicating

594. The posterior communicating arteries connect the internal carotid arteries with the __________________ arteries.

posterior cerebral

595. The internal carotid arteries subdivide into the _________ and _________ arteries.

anterior cerebral, middle cerebral

596. The right and left anterior cerebral arteries are connected through the _________ artery.

anterior connecting

745. From the left innominate vein the blood flows into the ________________ and then to the ________________ of the heart.

superior vena cava, right atrium

746. The blood flows to the bronchi through the bronchial arteries. The blood flows from the bronchi through the bronchial veins.

747. The blood flows to the muscles and skin of the chest wall through the intercostal arteries and from there through the intercostal veins.

748. From the bronchi the blood flows through the ________________ veins.

bronchial

749. From the muscles and skin of the chest wall the blood flows through the ______________ veins.

intercostal

750. The blood flows from the bronchial veins and from the intercostal veins through the azygos vein. (az' i gos)

751. The vein which receives the blood flowing from the intercostal veins is the ___________ vein.

azygos

752. The vein which receives the blood flowing from the bronchial veins is the ___________ vein.

azygos

753. The blood flows from the bronchi through the ________________ veins,

bronchial

754. into the __________________ vein.

azygos

755. The blood flows from the muscles and skin of the chest wall through the _____________ veins

intercostal

756. into the ________________ vein.

azygos

757. From the azygos vein the blood flows into the superior vena cava.

758. The superior __________________ receives the blood flowing from the azygos vein.

vena cava

759. From the azygos vein the blood flows into the _________________ and then to the right atrium of the heart.

superior vena cava

760. The blood flows to the bronchi through the ________________ arteries.

bronchial

597. Label the following diagram:

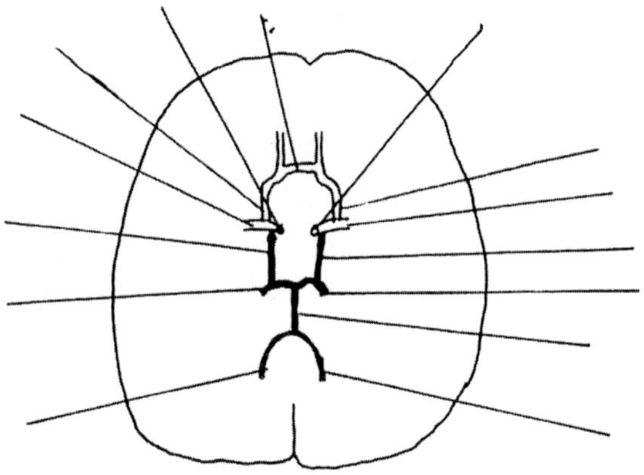

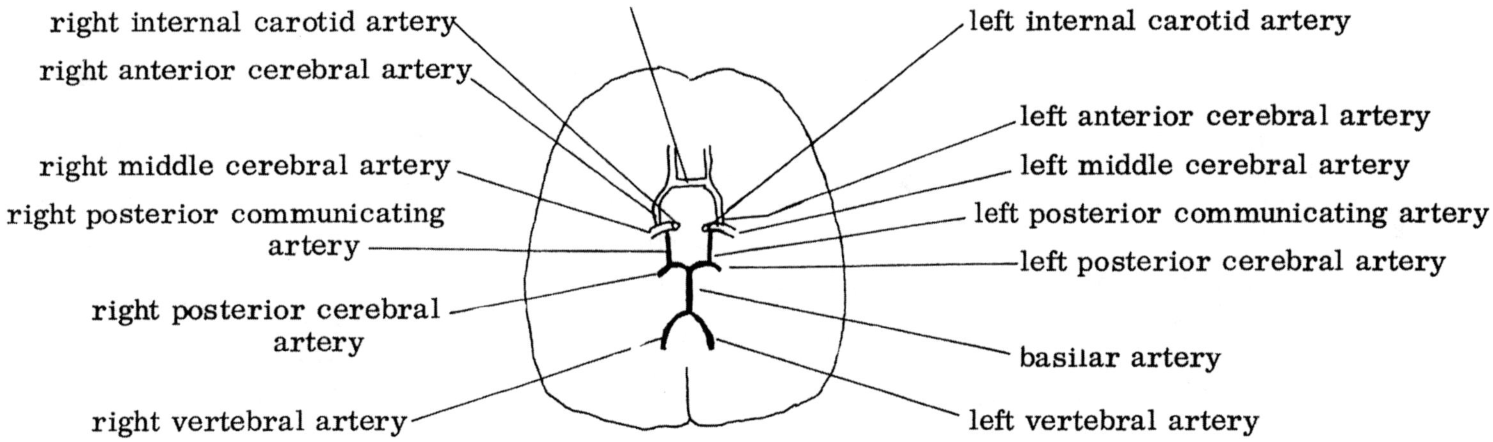

---

598. Which arteries comprise the circle of Willis?

The posterior cerebral arteries, the posterior communicating arteries, the anterior cerebral arteries, and the anterior communicating artery.

---

599. From the cerebral arteries the blood flows to all parts of the brain. From the brain the blood flows into the cerebral veins.

---

600. The veins which receive the blood flowing from the cerebral arteries of the brain are the ________________ veins.

---

cerebral

---

601. The blood flows to the brain through the cerebral arteries and from the brain through the ______________________.

---

cerebral veins

729. through the __________ of the aorta

arch

730. to the __________ aorta

thoracic

731. into the __________ arteries

superior phrenic

732. to the upper surface of the diaphragm and then into the __________ veins

superior phrenic

733. into the __________ vein

left innominate

734. into the __________ and return to the heart.

superior vena cava

735. The blood flows from the pericardium through the pericardial veins, and from the esophagus through the esophageal veins.

736. The veins carrying the blood from the esophagus are the __________ veins.

esophageal

737. The veins carrying blood from the pericardium are the __________ veins.

pericardial

738. The blood flows to the pericardium through the __________ arteries,

pericardial

739. and from the pericardium through the __________.

pericardial veins

740. The blood flows to the esophagus through the __________ arteries

esophageal

741. and from the esophagus through the __________.

esophageal veins

742. Both the esophageal veins and the pericardial veins empty into the left innominate vein.

743. The blood flowing from the pericardium flows into the __________ veins and then into the left __________ vein.

pericardial, innominate

744. The blood flowing from the esophagus flows into the __________ veins and then into the __________.

esophageal, left innominate vein

602. From the cerebral veins the blood flows into the various cranial venous sinuses. The uppermost of these is the superior sagittal sinus.

603. The location of the superior sagittal sinus is shown in the following diagram.

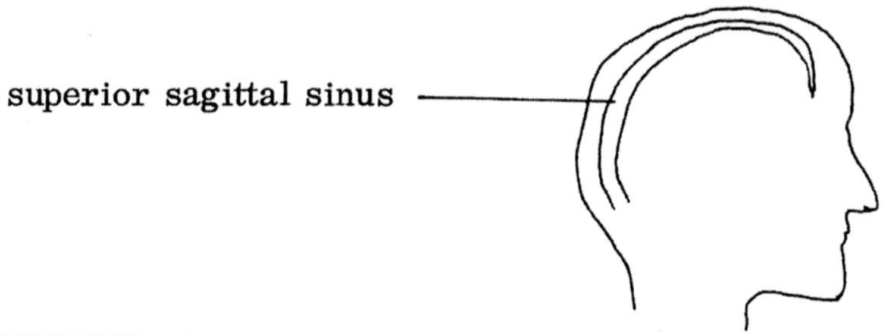

604. The sinus at the top of the brain is the ________________ sagittal sinus.

superior

605. The blood from the cerebral veins flows into the superior ______________ sinus.

sagittal

606. The sinus on the inner surface of the cranium is the ____________________ sinus.

superior sagittal

607. From the cerebral veins the blood flows into the __________________________.

superior sagittal sinus

608. The superior sagittal sinus connects with the straight sinus.

609. The sinus which connects with the superior sagittal sinus is the ______________ sinus.

straight

610. The superior sagittal sinus connects with the ________________ sinus.

straight

611. Label the following diagram:

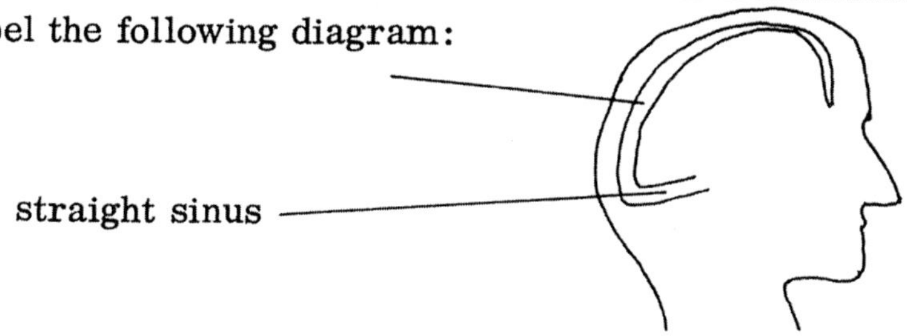

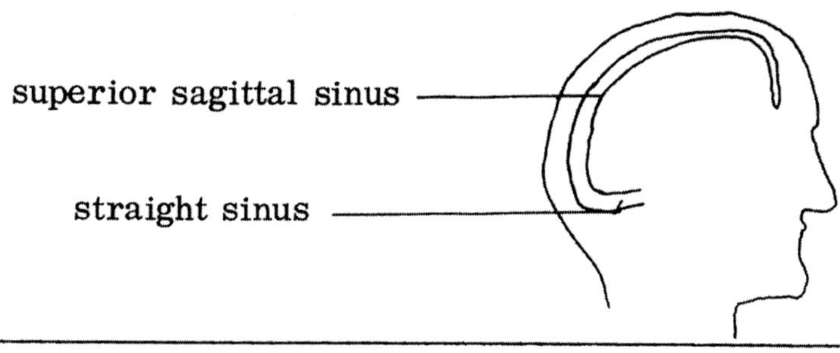

714. The _______________ arteries supply blood to the pericardium.

pericardial

715. The _______________ arteries supply blood to the esophagus.

esophageal

716. The _______________ arteries supply blood to the muscles and skin of the chest wall.

intercostal

717. The _______________ arteries supply blood to the upper surface of the diaphragm.

superior phrenic

718. The blood flows from the thoracic aorta through the visceral and parietal arteries, through the arterioles and capillaries to the body parts and then into the venules and veins.

719. From the upper surface of the diaphragm the blood flows into the superior phrenic veins.

720. The blood flows from the upper surface of the diaphragm through the superior _______________ veins.

phrenic

721. The blood flows to the upper surface of the diaphragm through the _______________ arteries.

superior phrenic

722. The blood flows from the upper surface of the diaphragm through the ____________ veins.

superior phrenic

723. From the superior phrenic veins the blood flows into the left innominate vein.

724. The ___________ innominate vein receives the blood flowing from the superior phrenic veins.

left

725. From the upper surface of the diaphragm the blood flows into the _______________ veins and then into the left _______________ vein.

superior phrenic, innominate

726. The ____________ vein receives the blood flowing from the superior phrenic veins.

left innominate

727. The left innominate vein empties into the _______________ which carries the blood to the right atrium.

superior vena cava

728. Following the path of the blood from the heart to the upper surface of the diaphragm and return, the blood leaves the heart and flows into the _______________ aorta,

ascending

612. The superior sagittal sinus connects with the straight sinus to form the paired (right and left) transverse sinuses.

613. When the superior sagittal sinus joins the straight sinus they form the paired ______________ sinuses.

transverse

614. The blood from the brain flows into the superior sagittal sinus which joins the straight sinus to form the paired ______________.

transverse sinuses.

615. Label the following diagram:

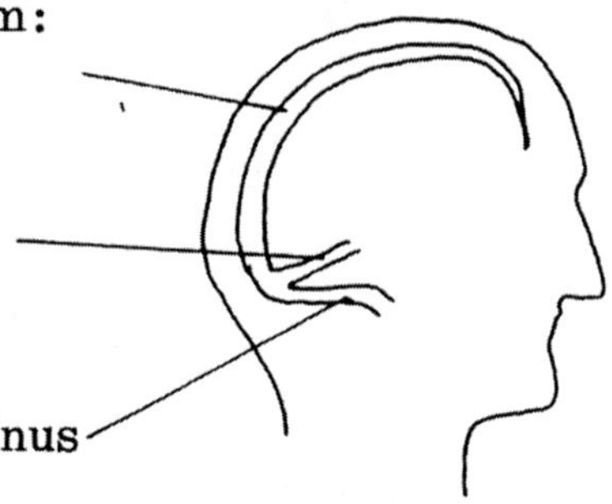

transverse sinus

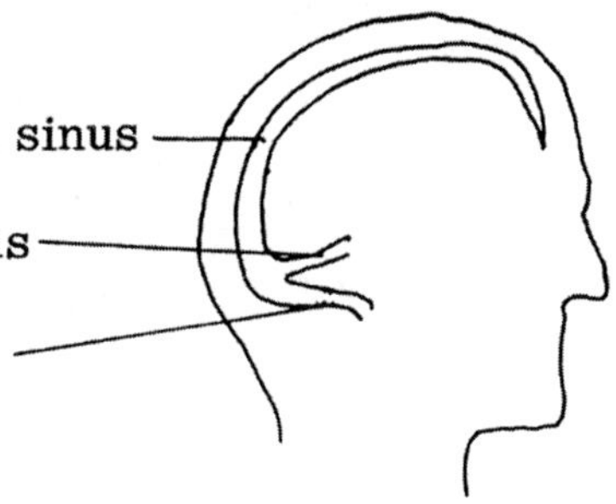

superior sagittal sinus

straight sinus

transverse sinus

616. Leading into the straight sinus is the inferior sagittal sinus.

617. The sinus which leads into the straight sinus is the ______________ sagittal sinus.

inferior

618. The blood flowing into the inferior ______________ sinus enters the straight sinus.

sagittal

619. The superior sagittal sinus connects with the straight sinus to form the paired transverse sinuses. Leading to the straight sinus is the ______________ sinus.

inferior sagittal

701. The intercostal arteries lead to the skin and muscles of the chest wall while the superior phrenic arteries lead to the upper surface of the diaphragm.

---

702. The arteries which lead to the skin and muscles of the chest wall are the ____________**costal** **arteries**.

inter

---

703. The parietal arteries leading to the muscles and skin of the chest wall are the ____________ arteries.

intercostal

---

704. The parietal arteries leading from the thoracic aorta are the ______________ arteries and the superior phrenic arteries.

intercostal

---

705. The parietal arteries leading from the thoracic aorta are the intercostal arteries and the ______________ phrenic arteries.

superior

---

706. The parietal arteries leading to the upper surface of the diaphragm are the superior ______________ arteries.

phrenic

---

707. The ______________ arteries supply blood to the upper surface of the diaphragm.

superior phrenic

---

708. The ______________ arteries supply blood to the muscles and skin of the chest wall.

intercostal

---

709. The parietal branches of the thoracic aorta are the ______________ and ______________ arteries.

intercostal, superior phrenic

---

710. The thoracic aorta gives off many small branches which may be divided into two groups, the ______________ arteries and the ______________ arteries.

visceral, parietal

---

711. The visceral branches of the thoracic aorta are the ______________, the ______________ and the ______________ arteries.

bronchial, esophageal, pericardial

---

712. The parietal branches of the thoracic aorta are the ______________ and the ______________ arteries.

intercostal, superior phrenic

---

713. The ______________ arteries supply blood to the bronchi.

bronchial

---

620. Label the following diagram:

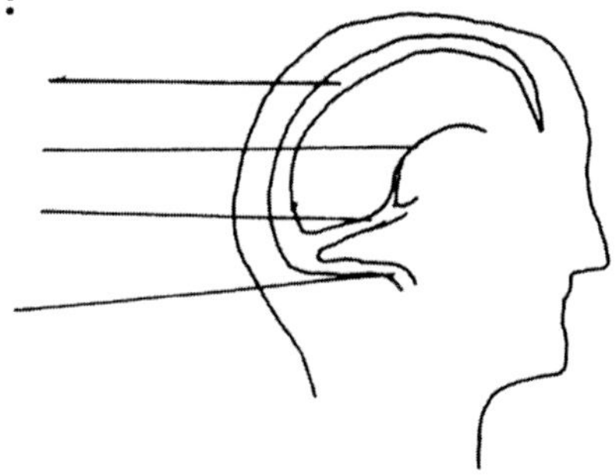

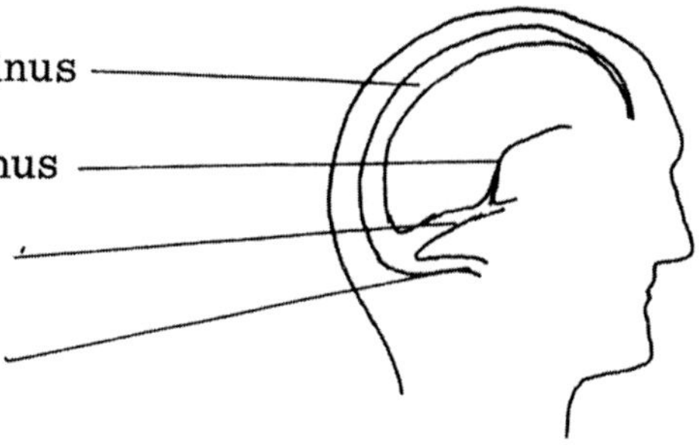

---

621. The paired transverse sinuses lead into the right and left internal jugular veins.

---

622. The right and left internal ______________ veins receive the blood flowing from the various cranial venous sinuses. They also receive the blood flowing from the veins of the face and neck.

jugular

---

623. From the paired transverse sinuses the blood flows into the right and left ______________ jugular veins.

internal

---

624. The veins which receive the blood flowing from the transverse sinuses are the right and left ______________ veins.

internal jugular

---

625. From the paired transverse sinuses the blood flows into the ______________.

right and left internal jugular veins

689. The branches of the thoracic aorta extending to the esophagus are the __________ arteries.

esophageal

690. The branches of the thoracic aorta extending to the pericardium are the __________ arteries.

pericardial

691. The __________ arteries extend to the bronchi.

bronchial

692. The visceral branches of the thoracic aorta are the bronchial arteries which extend to the bronchi, the pericardial arteries which extend to the pericardium, and the __________ arteries which extend to the esophagus.

esophageal

693. Three visceral branches of the thoracic aorta are the bronchial arteries, the pericardial arteries, and the __________ arteries.

esophageal

694. The pericardial arteries, the esophageal arteries, and the __________ arteries make up the visceral branches of the thoracic aorta.

bronchial

695. The visceral branches of the thoracic aorta are the __________ arteries, the __________ arteries, and the pericardial arteries.

bronchial, esophageal

696. From the thoracic aorta the visceral branches lead to the corresponding organs. The __________ arteries lead to the bronchi, the __________ arteries to the esophagus, and the __________ arteries to the pericardium.

bronchial, esophageal, pericardial

697. The visceral branches of the thoracic aorta are the __________, __________, and __________ arteries.

bronchial, esophageal, pericardial

698. From the thoracic aorta the visceral arteries extend to the chest organs.
The __________ arteries extend to the __________;
the __________ arteries to the __________; and
the __________ arteries to the __________.

bronchial, bronchi
esophageal, esophagus
pericardial, pericardium

699. The thoracic aorta has many small arteries branching from it. These may be divided into two main groups, the visceral arteries and the __________ arteries.

parietal

700. The parietal arteries branching from the thoracic aorta are the intercostal arteries and the superior phrenic arteries.

626. Label the following diagram:

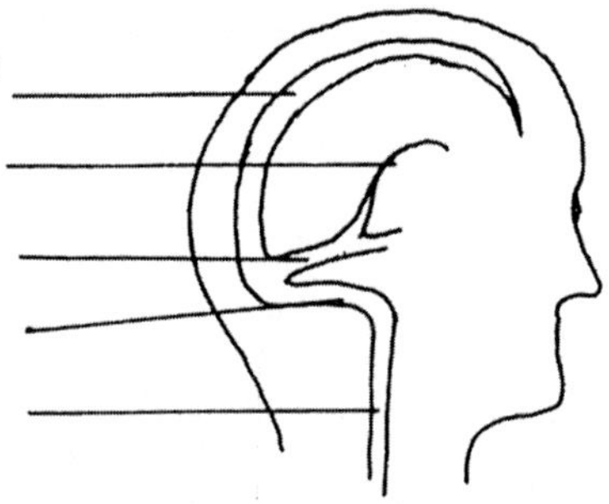

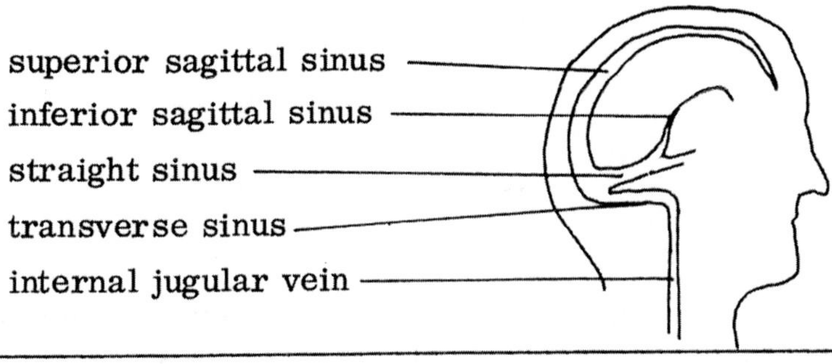

---

627. The right internal jugular vein joins the right subclavian vein to form the right innominate vein. The left internal jugular vein joins the left subclavian vein to form the left __________ vein.

innominate

---

628. The internal jugular veins join the corresponding subclavian veins to form the right and left ____________________ veins.

innominate

---

629. The blood flows from the right and left internal jugular veins to join the right and left ____________________ veins

subclavian

---

630. which then form the right and left ____________________ veins.

innominate

---

631. The right and left innominate veins lead into the ________________ vena cava which in turn leads into the ____________________ of the heart.

superior, right atrium

---

632. The blood from the cerebral arteries flows into the cerebral veins and then into the cranial venous sinuses. The sinus at the top of the brain is the ________________________ sinus.

superior sagittal

---

633. The superior sagittal sinus joins the _______________________ sinus

straight

---

634. to form the paired ___________________ sinuses.

transverse

VII. THE THORACIC AORTA

678. As the aorta leaves the heart, the upward part is called the ascending aorta. The curved part is called the arch of the aorta and the descending part, down to the diaphragm, is called the thoracic aorta.

---

679. Label:

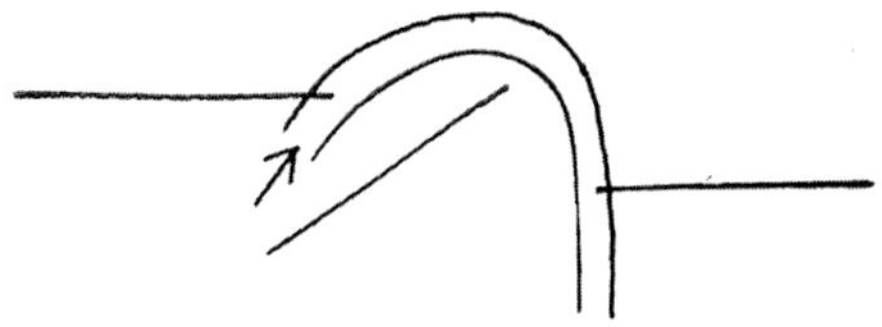

- -

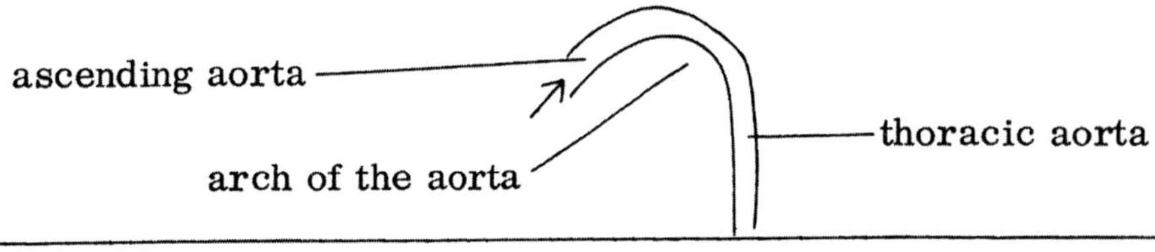

---

680. The thoracic aorta has many small branches which may be divided into two main groups, the visceral arteries which supply blood to the viscera of the chest and the parietal arteries which supply blood to the walls of the chest cavity.

---

681. The visceral arteries supply blood to the ____________ while the parietal arteries supply blood to the walls of the chest cavity.

- -

viscera

---

682. The arteries which supply blood to the viscera of the chest are called the __________ arteries.

- -

visceral

---

683. The parietal arteries supply blood to the walls of the ____________________.

- -

chest cavity

---

684. In the chest the visceral arteries supply blood to the viscera while the _______________ arteries supply blood to the chest wall.

- -

parietal

---

685. The branches of the thoracic aorta may be divided into two main types, the ______________ arteries and the ______________ arteries.

- -

visceral, parietal

---

686. The visceral arteries branching from the thoracic aorta are the bronchial arteries, the esophageal arteries, and the pericardial arteries. Each branch extends to the structure named.

---

687. The bronchial arteries extend to the bronchi and the esophageal arteries extend to the ______________.

- -

esophagus

---

688. Likewise the pericardial arteries extend to the ____________________.

- -

pericardium

---

635. Leading into the straight sinus is the ________________sinus.

inferior sagittal

636. The paired transverse sinuses lead into the ______________________

right and left internal jugular veins

637. which connect with the corresponding __________________ veins

subclavian

638. to form the right and left ____________________ veins

innominate

639. which lead into the ________________

superior vena cava

640. and then to the ________________ of the heart.

right atrium

641. Label:

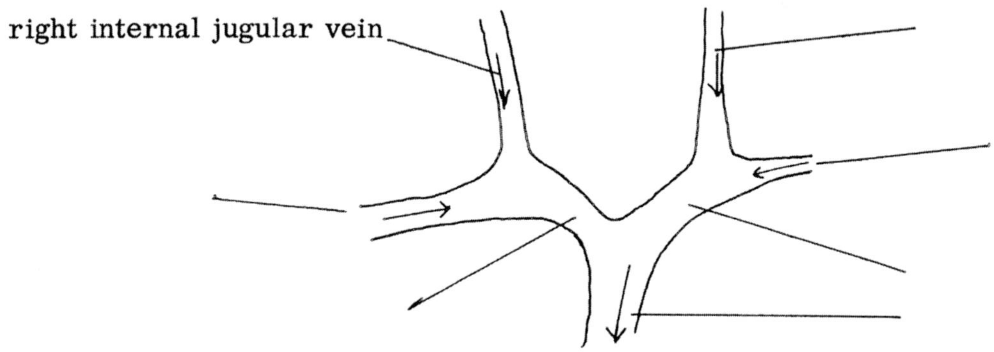

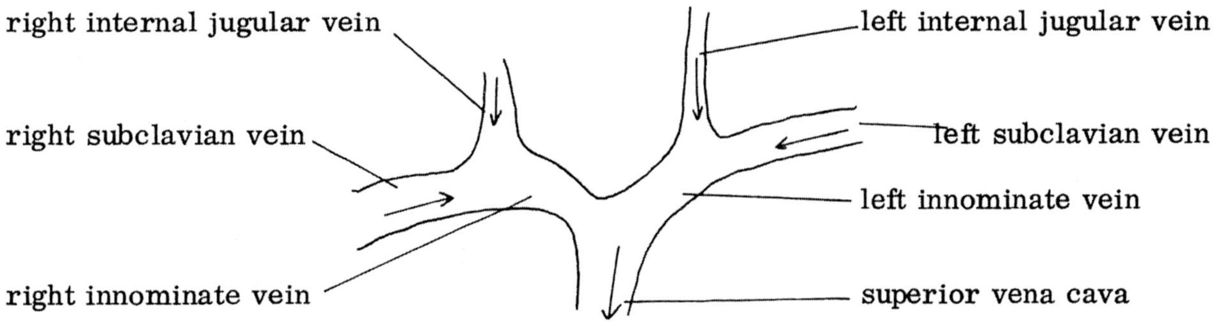

671. Connecting with the straight sinus is the ________________ sinus.

inferior sagittal

672. From the paired transverse sinuses the blood flows into the right and left ____________ veins

internal jugular

673. to meet the right and left ________________ veins

subclavian

674. and then form the right and left ________________ veins

innominate

675. which lead into the ________________

superior vena cava

676. and back to the ________________ of the heart.

right atrium

677. Trace the path of the blood from the right atrium to the brain and return.

Right atrium through tricuspid valve to right ventricle through semilunar valves to pulmonary artery to lungs to pick up oxygen and give off carbon dioxide to pulmonary veins to left atrium through bicuspid valve to left ventricle through semilunar valves to ascending aorta to arch of aorta. From arch of aorta to innominate artery which subdivides into right common carotid artery and right subclavian artery. Right subclavian artery which extends upward through the vertebral artery. From arch of aorta through left common carotid artery and left subclavian artery. From left subclavian artery to left vertebral artery. Right and left vertebral arteries to base of brain to form basilar artery to right and left posterior cerebral arteries. Right and left common carotid arteries to right and left internal and external carotid arteries. Right and left internal carotid arteries (branch to the eyes, the ophthalmic arteries) to base of brain to posterior communicating arteries which connect to posterior cerebral arteries and then to anterior cerebral and middle cerebral arteries. Right and left anterior cerebral arteries connected by anterior communicating artery. Then to cerebral veins to superior sagittal sinus to straight sinus and transverse sinuses (with inferior sagittal sinus connecting to straight sinus) to right and left internal jugular veins to join right and left subclavian veins to form right and left innominate veins to superior vena cava to right atrium.

642. Label:

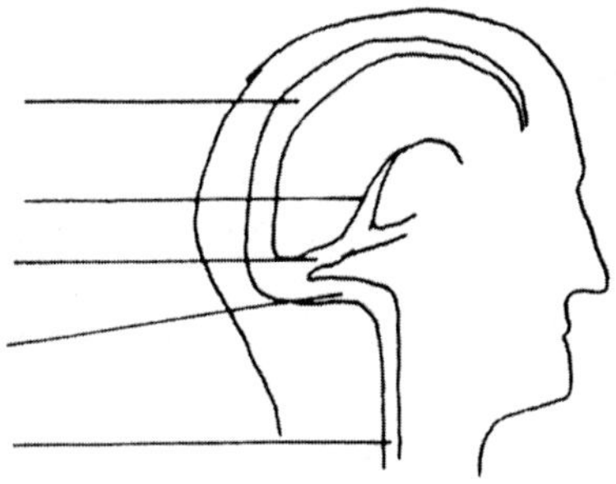

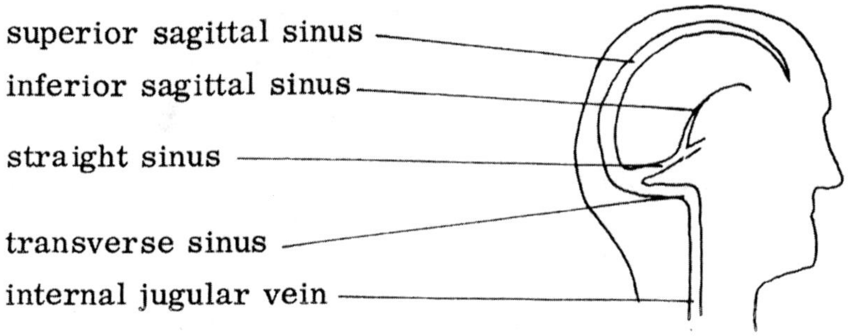

643. Label:

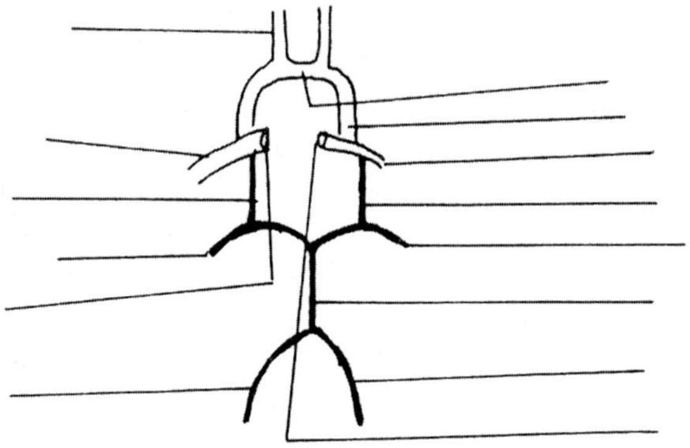

base of brain

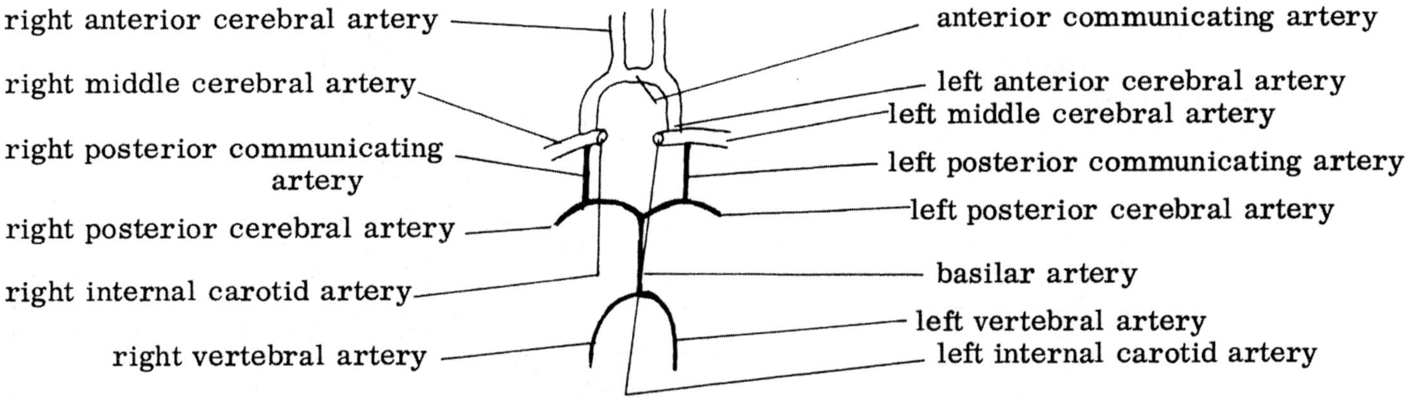

658. The two other branches from the arch of the aorta are the left ________________ artery and the left ________________ artery.

subclavian, common carotid

659. The upward branch extending from the left subclavian artery is the ____________________ artery.

left vertebral

660. The right and left vertebral arteries extend upwards to the base of the brain where they join to form the ______________ artery.

basilar

661. The basilar artery subdivides into the right and left _____________________ arteries.

posterior cerebral

662. The right and left common carotid arteries extend upwards where they divide into the right and left _____________ and _____________ arteries.

internal carotid, external carotid

663. The right and left internal carotid arteries send one branch, the _______________ artery to the eyes.

ophthalmic

664. At the base of the brain the internal carotid arteries give off branches to the posterior cerebral arteries. These branches are called the _____________________ arteries.

posterior communicating

665. The internal carotid arteries then subdivide into the _________________ arteries and the _______________ arteries.

anterior cerebral, middle cerebral

666. The right and left anterior cerebral arteries are connected by the _________________ artery.

anterior communicating

667. These interconnecting arteries form the Circle of Willis.

668. From the cerebral arteries the blood flows into the cerebral veins and then into the cranial venous sinuses. The uppermost venous sinus is the ___________________ sinus,

superior sagittal

669. which connects with the ________________ sinus

straight

670. to form the paired ______________________ sinuses.

transverse

644. Time for a review from the very beginning. (If you know the entire circulation through the heart to the brain and return, go on to frame 677).

The circulation from the right atrium of the heart to the brain and return is as follows:

Blood enters the heart at the right atrium and then flows through the ______________ valve

---

tricuspid

645. into the ______________.

---

right ventricle

646. From the right ventricle the blood passes through the ______________ valves

---

semilunar

647. into the ______________ artery

---

pulmonary

648. and then to the lungs where it picks up ______________ and gives off ______________.

---

oxygen, carbon dioxide

649. From the lungs the blood flows into the ______________ veins

---

pulmonary

650. to the ______________ of the heart.

---

left atrium

651. From the left atrium the blood flows through the ______________ valve

---

bicuspid

652. into the ______________.

---

left ventricle

653. From the left ventricle the blood flows through the ______________ valves

---

semilunar

654. into the ______________ aorta.

---

ascending

655. From the ascending aorta the blood flows into the arch of the aorta where it goes into right and left paths. The right branch is called the ______________ artery

---

innominate

656. which in turn branches into the right ______________ artery and the right ______________ artery.

---

subclavian, common carotid

657. From the right subclavian artery, the branch extending upward is called the ______________ artery.

---

right vertebral

Turn the book around and start back toward the front of the book.